the Eleme...

							NOBLE GASES	
IB	**IIB**	**IIIA**	**IVA**	**VA**	**VIA**	**VIIA**		
						1 **H** 1.0079 †	2 **He** 4.00260	2
		5 **B** 10.81	6 **C** 12.011	7 **N** 14.0067	8 **O** 15.9994†	9 **F** 18.99840	10 **Ne** 20.179†	2 8
		13 **Al** 26.98154	14 **Si** 28.086 †	15 **P** 30.97376	16 **S** 32.06	17 **Cl** 35.453	18 **Ar** 39.948†	2 8 8
29 **Cu** 63.546†	30 **Zn** 65.38	31 **Ga** 69.72	32 **Ge** 72.59 †	33 **As** 74.9216	34 **Se** 78.96 †	35 **Br** 79.904	36 **Kr** 83.80	2 8 18 8
47 **Ag** 107.868	48 **Cd** 112.40	49 **In** 114.82	50 **Sn** 118.69 †	51 **Sb** 121.75†	52 **Te** 127.60†	53 **I** 126.9045	54 **Xe** 131.30	2 8 18 18 8
79 **Au** 196.9665	80 **Hg** 200.59†	81 **Tl** 204.37†	82 **Pb** 207.2	83 **Bi** 208.9804	84 **Po** (210)	85 **At** (210)	86 **Rn** (222)	2 8 18 32 18 8

	64 **Gd** 157.25†	65 **Tb** 158.9254	66 **Dy** 162.50†	67 **Ho** 164.9304	68 **Er** 167.26†	69 **Tm** 168.9342	70 **Yb** 173.04†	71 **Lu** 174.97	2 8 18 32 9 2
96 **Cm** (247)	97 **Bk** (247)	98 **Cf** (251)	99 **Es** (254)	100 **Fm** (257)	101 **Md** (258)	102 **No** (255)	103 **Lr** (256)		2 8 18 32 32 9 2

$7.95

TRACE ELEMENTS AND MAN:
Some Positive and Negative Aspects

by HENRY A. SCHROEDER, M.D.

It is Dr. Schroeder's thesis that life originated in the sea, and when animals—including man—became terrestrial, they brought with them a necessity for those trace elements which occur naturally in sea water. These elements, generally present in minute amounts, are essential for normal metabolism. They include chromium, iodine, copper, cobalt, zinc, manganese, and a few others which are discussed in this breakthrough book.

Dr. Schroeder further contends that certain other elements, "locked in" to the earth's crust for the most part, have in recent years been released by industrial man into the atmosphere and the environment generally. These—cadmium, lead and mercury especially—are having an adverse effect on human health and human life.

Unlike vitamins, which are synthesized plants and animals, minerals—of equal or even greater importance—must come from outside; and failing to provide them in our diets can lead to trouble. Chromium deficiency, for instance, is revealed as a prime cause of the nation's number one killer: atherosclerosis. A faulty diet fails to provide this, as well as other essential trace elements.

The fastest way to develop atherosclerosis, Dr. Schroeder suggests, is to drink coffee all day with three or four spoonfuls of sugar, plus cream; use marmalades and jams on breakfast toast, with white sugar thickly sprinkled on processed breakfast foods; eat a couple of ham sandwiches on white bread for lunch, followed by a slab of pie; and for dinner, large portions of pork with rich gravies, deep-fried potatoes, and a lemon meringue pie for dessert.

Dr. Schroeder has become a world authority in the little known field of the trace elements as they relate to living organisms. His findings are the result of years of experimenting at the unique Dartmouth Medical School Trace Element Laboratory in Brattleboro, Vermont. He is the author of seven books, including *A Matter of Choice*, and *Pollution, Profits and Progress*.

THE DEVIN-ADAIR COMPANY, Publishers
Old Greenwich, Connecticut

THE TRACE ELEMENTS AND MAN

*"For there is nothing covered, that shall not be revealed;
Neither hid, that shall not be known."*

ST. LUKE XII, 2.

THE TRACE ELEMENTS
AND MAN

Some Positive and Negative Aspects

by

Henry A. Schroeder, M.D.

THE DEVIN-ADAIR COMPANY
Old Greenwich, Connecticut

Library of Congress Card Number:
72-85732

ISBN 0–8159–6907–4

Second printing, 1975

Printed in the United States of America

To MARIAN MITCHENER,

whose care and devotion to mice and rats

made this story possible.

Preface

THIS BOOK WAS WRITTEN to fulfill an obligation. When the people of a democratic society support scientific research through their taxes, they have a right to expect full publication of the results they paid for. When the people support biological and medical research they also have a right to hope that some increment of knowledge of the processes of life or the prevention or cure of a disease may be discovered. If nothing is discovered, as happens occasionally, the people have the right to expect to be told about it.

Food is one of the greatest concerns of the informed American. It has been well said that a man digs his grave with his teeth. The public is health-conscious and is convinced that food and health are so closely related as to be interdependent. The need for a "balanced" diet has been drummed for years into the public consciousness by nutritionists. (Because natural foods are naturally "balanced," the need to balance the diet does not exist in any living creature other than man and his animals.) There are "crash" diets which are not balanced, and curious perversions of diets to reduce weight faster than it should be lost, made up of all of the combinations of foods possible: all carbohydrate—protein and alcohol-protein and fat without carbohydrate—high carbohydrate, low fat and protein diets—vegetarian diets—all meat diets—and other strange and weird regimens. Each of them neglects the fact that the human animal evolved as an omnivore. It is a tribute to the marvelous economy and adaptability of the human body that it can stand some of these diets without more than temporary harm. There is no magic in any of them.

There is a sizeable segment of the public which is convinced that the mass production of food and its necessary refinement, processing, and storage add subtle toxins to food and remove some essential factors. Although the average level of nutrition in the United States and in other civilized countries has never been higher, the population, especially older people, have many diseases caused directly or indirectly by food. There is some truth in these beliefs.

Certain changes in the natural environment of man have appeared as a by-product of modern civilization. Among them are an increased exposure to certain elements unnecessary for or even toxic to health, most of which are mined from the earth and get into food, water and air. In other cases, certain micronutrients necessary for health and even for life itself have been partly removed during the refining of foods. Because the trace elements are so vital for health, this book explores the possibility that diseases may result from deficiencies or excesses.

The trace elements are more important in nutrition than are the vitamins, in the sense that they cannot be synthesized by living matter, as is the case with vitamins. Thus they are the basic spark-plugs in the chemistry of life, on which the exchanges of energy in the combustion of foods and the building of living tissues depend. Every natural food contains the micronutrients necessary for its metabolism. It is only when natural foods are altered by man—cooked, partitioned, preserved, frozen, canned, dried, processed or packed—that they lose or gain micronutrients or toxins which can affect health over the long term.

How much do we require? How much of each do we get? How much do domestic animals require and get? What is the function of each? What about the "new" trace elements, introduced in large quantities into Man's environment in ever increasing amounts since the Bronze Age? Are any of them harmful in some subtle manner to civilized people? If so, what harm do they do? These questions are answered in so far as possible.

I am indebted to the scientists, few in number, who have devoted their interests to studies of trace elements, and especially to those who have pioneered in the subject. Among them are Isabel H. Tipton (all elements), Harriet L. Hardy (beryllium), George L. Curran (vanadium and chromium), Walter Mertz (chromium), George C. Cotzias (manganese), I. Herbert Scheinberg (copper), Bert L. Vallee (zinc), E. J. Underwood (animal nutrition), G. Monier Williams (elements in food), Clair C. Patterson (lead) and G. Parry Howells (who provided me with her exhaustive survey of all elements, intake, metabolism and excretion, in Standard Man, from which much of the data have been abstracted). I am particularly indebted to our supporters, the National Institutes of Health, through whom so many tax dollars flow.

HENRY A. SCHROEDER, M.D.

Brattleboro, Vermont and
St. Thomas, Virgin Islands

CONTENTS

CHAPTER I

The Periodic Table
of the Elements

A hundred years ago, Dmitri I. Mendeléyeff, a Russian chemist was studying the weights of all the known elements previously discovered on this earth. Knowing something of their properties in solution, he listed them in order, for Mendeléyeff believed in Natural Order. In a sudden flash of inspiration, he detected a periodicity of property and weight. After several trials—in fact many, for the finished product was the result of a number of revisions—he drew up a table of these periodic repetitions, which followed what he called "Natural Law." Because Mendeléyeff was a true genius believing in order, he left blank spaces where he was convinced an element should be, for he was sure that just because an element had not been discovered, there was no reason to think that it could not exist. To Mendeléyeff came a revelation that was the most important single discovery of modern science, for it opened the fields of chemistry and physics to rational investigation and proved the beautifully simple order of matter in the Universe and the structure of the atom. (See Periodic Chart of the Elements, endpaper.)

Each of Mendeléyeff's blank spaces has since been filled with new elements having properties predicted by him, from atomic numbers 1 through 92, several of them during his life time. They were found because he believed that they should be there, or there was no true order in the Universe. Furthermore, with the burgeoning of atomic physics, the reasons why the elements fell into a periodic order were discovered in the nature and structure of electrons, or negative charges, protons or positive charges, and neutrons having no charge, which made each atom different. The radioactive elements, some of which have short lives, fell into place. In table I-1 are shown their order in terms of weight, structure and chemical properties.

With the stimulus to research resulting from the discovery of fission leading to the atomic bomb, brand new elements were made. Mendeléyeff had left eleven blank spaces in the actinium series after uranium, No. 92; all have since been filled, from 93 and 94, neptunium and plutonium, to the last, lawrencium, No. 103. Whether or not man can go further is problematical, for he has to begin a new series, and most of the others have very short existences, being very unstable—which is why they do not occur naturally.

As this book is concerned with life and living things, we do not need to consider radioactive elements unless they accumulate in living things naturally or from man-made exposures. All of the elements which are needed by living things are contained in the first 53 of the 92 natural elements on the surface of the earth; all but one occurs in the first 42 and all but two in the first 34.

In the Universe there is a natural order of abundance of the elements, following atomic weight and number. The heavier the element, the less abundant it is. Furthermore, elements with even atomic numbers are more abundant than elements with odd atomic numbers. This order may not be true now for hydrogen and helium, but it will be eventually in the Sun and other stars. Four hydrogen atoms fuse to make one helium atom, releasing a tremendous amount of energy in the form of light and heat, and one can measure the age of the sun by knowing the proportions of hydrogen and helium. The "hydrogen," or fusion bomb, is based on this reaction which is natural in the stars.

No one knows why odd numbered elements are less abundant than even numbered ones. Until we know more of the

Master Plan of matter in the Universe we cannot even guess. Nor do we know why the general order of abundance of elements in the Universe according to atomic number is not wholly true on the earth. If it held, lithium and beryllium would be the most abundant solid matter on earth. All we can guess is that a planet is not wholly representative of the Universe, which we know, and that perhaps these light metals were not formed as readily as heavier ones. But in general, on the earth the order of relative abundance of the elements is the rule, with a number of exceptions, including, of course, the so-called "inert gases" (which are no longer inert, being made to react with fluorine under special non-universal conditions).

For an element to take part in living matter, it must be: a) abundant where life began, in sea water; b) reactive; that is, able to join up or bond with other elements: c) able to form an integral part of a structure, and d) in the case of the metals, soluble in water, reactive of itself with oxygen, and able to bond to organic material (carbon, hydrogen, oxygen, nitrogen, sulfur, phosphorus). These qualities are found in 23 elements, and perhaps 27, of the first 42 of the Periodic Table.

When one looks at the Periodic Table, as I have for some 20 years, one is constantly discovering new bits of information. Look at it horizontally, especially at the elements with numbers 22 to 30. This is called the first transitional series. Each atom has one or more electrons missing from its outer ring. An atom is like a solar system, the electrons being the planets, which revolve about the central mass of protons at enormous speed. As one goes from titanium, No. 22, to zinc, No. 30, the unfilled orbits change, one electron at a time, changing the properties of each metal in respect to its reactivity. Zinc's outer ring is filled. This transition is similar in Nos. 40-48, and in Nos. 72-80.

Look at the table vertically. Each element down from the top row is like the one above it, only heavier. Its outer electron ring is exactly like that of the one above, only larger. Therefore, it has the same chemical and physical properties, only a bit different because of size. Its reactions are the same. It is likely to occur, and does occur, in nature with the one above. Copper,

TABLE I-1

Natural elements of the Periodic Table in the Universe, excluding the rare earths.

Atomic Number	Symbol		State	Atomic Number	Symbol		State
1	H	Hydrogen	G	37	Rb	Rubidium	M
2	He	Helium	G	38	Sr	Strontium	M
3	Li	Lithium	M	39	Y	Yttrium	M
4	Be	Beryllium	M	40	Zr	Zirconium	TM
5	B	Boron	C	41	Nb	Niobium	TM
6	C	Carbon	C	42	Mo	Molybdenum	TM
7	N	Nitrogen	G	43	Tc	Technetium	TMR
8	O	Oxygen	G	44	Ru	Ruthenium	TM
9	F	Fluorine	C	45	Rh	Rhodium	TM
10	N	Neon	G	46	Pa	Palladium	TM
11	Na	Sodium	M	47	Ag	Silver	TM
12	Mg	Magnesium	M	48	Cd	Cadmium	M
13	Al	Aluminum	M	49	In	Indium	M
14	Si	Silicon	C	50	Sn	Tin	M
15	P	Phosphorus	C	51	Sb	Antimony	M
16	S	Sulphur	C	52	Te	Tellurium	C
17	Cl	Chlorine	C	53	I	Iodine	C
18	Ar	Argon	G	54	Xe	Xenon	G
19	K	Potassium	M	55	Cs	Cesium	M
20	Ca	Calcium	M	56	Ba	Barium	M

No.	Symbol	Name	Type
21	Sc	Scandium	M
22	Ti	Titanium	TM
23	V	Vanadium	TM
24	Cr	Chromium	TM
25	Mn	Manganese	TM
26	Fe	Iron	TM
27	Co	Cobalt	TM
28	Ni	Nickel	TM
29	Cu	Copper	TM
30	Zn	Zinc	M
31	Ga	Gallium	M
32	Ge	Germanium	M
33	As	Arsenic	C
34	Se	Selenium	C
35	Br	Bromine	C
36	K	Krypton	G
57	La	Lanthanum	M
72	Hf	Hafnium	TM
73	Ta	Tantalum	TM
74	W	Wolfram(Tungsten)	TM
75	Re	Rhenium	TM
76	Os	Osmium	TM
77	Ir	Iridium	TM
78	Pt	Platinum	TM
79	Au	Gold	TM
80	Hg	Mercury	M
81	Tl	Thallium	M
82	Pb	Lead	M
83	Bi	Bismuth	M
84	Po	Polonium	MR
90	Th	Thorium	MR
92	U	Uranium	M

G Gas. The "inert" gases, which are only relatively inert, are shown to the right of the Table. In their molecular, or uncombined states, fluorine and chlorine are gases, but they always occur naturally as compounds.

M Metal. Metals in Group 1A are alkali metals, in IIA, alkaline earths. They always occur as compounds in nature, with a few exceptions, such as silver, gold, mercury.

C Non-Metals occurring as compounds, often with oxygen, Group VIIA as simple salts of sodium.

TM Transitional Metal with unfilled outer electron shell, hence reactive.

R Radioactive, hence unstable to some degree. Elements numbered 84-92 are radioactive: Polonium, Astatine, Radon, Francium, Radium, Actinium, Thorium, Protactinium, Uranium, the last having little radioactivity in its natural state. Most of them are not listed.

silver and gold occur in the same ores, so do zinc and cadmium, nickel, palladium and platinum, vanadium, niobium and tantalum, chromium, molybdenum and tungsten (wolfram) and so on.

This property is important biologically, especially in terms of reactions and in the possibility of disease. For sizeable amounts of a heavier metal can displace a lighter one in the same group in biological tissues and alter the reaction of the lighter one. Furthermore, when tissues have an affinity for a certain element or are structured by it, they have an affinity for all other elements of the group. Thus, all Group IIA elements are bone seekers, all Group VII elements are thyroid seekers, all Group IIB and VIB are liver and kidney seekers. There seem to be specialized organic compounds in these organs which are avid for certain kinds of elements. Similar compounds have been manufactured in large varieties by chemists; they are called chelating agents, from the Greek *chela,* a claw. We do not know what these tissue compounds are, but we do find groups of metals in special tissues, particularly when exposures to the "abnormal" or unnatural elements are heavy.

We can think of the biological system in simple terms as a set of two intermeshing gears, one an organic compound, usually a protein and usually an enzyme, or catalyst, which initiates or speeds up chemical reactions at body temperatures—reactions which might otherwise not begin. (Sugar and fat are burned at body temperatures, using oxygen and giving up carbon dioxide and water, just as wood and oil are burned at much higher temperatures. These catalysts, which start all biological chemical reactions, make substances burn at low temperatures in water.) The other gear which makes the system go is a metal, which we will say, meshes exactly with the organic gear.

Now let us change the metal gear to a larger one with the same sized teeth. Obviously, the system will go faster. It may go so fast that it may break. At any rate, it changes, and will not do what it is supposed to do. Again, let us put in a gear twice the size of the second. It does not fit at all, so the small gear stays. This somewhat far-fetched explanation illustrates how a metal exactly like the necessary one, only a bit larger, can interfere with an orderly system and cause disorder. Thus, pairs of metals in the same periodic group can interact, especially if there are more of the larger than the smaller.

The analogy of a key in a lock is also applicable. The correct key fits the keyhole and unlocks. The same make of key will fit the keyhole but will not unlock. A different make of key will not fit the keyhole.

In this way, niobium can displace vanadium, tungsten can displace molybdenum (we do not know about chromium and molybdenum), technetium could displace manganese (if one could obtain it in quantities, but it is radioactive and short-lived); ruthenium could displace iron, rhodium cobalt, palladium nickel; silver does displace copper, gold will displace copper under certain circumstances, cadmium avidly displaces zinc and changes or inactivates zinc enzymes, causing disease; arsenic displaces phosphorus, causing disease; selenium displaces sulfur, causing disease; bromine displaces chlorine; beryllium displaces magnesium; magnesium and calcium interact; strontium displaces calcium; lithium displaces sodium; sodium and potassium interact; rubidium displaces potassium, as does cesium in bacteria, and chemically, silicon can substitute for carbon making silicon rubber, or "silly putty," a fascinating plaything which has no conceivable use.

Somewhere on a distant and very hot planet there may be a form of life using silicon, phosphorus, sulfur and fluorine instead of carbon, nitrogen, oxygen and hydrogen. Personally, I doubt it, for reactions would be sluggish and other necessary elements would be gaseous. When we go deep enough into the properties of the elements and their places and functions in living things, we cannot fail to become convinced that some of the elements were designed for the purpose of making combinations which could live, and the way they interact is the only way in which life could exist in the Universe. For all matter in the Universe is made up of elements in the Periodic Table, revealed by the genius of Mendeléyeff a hundred years ago, in 1871.

CHAPTER II

The Beginnings of Life and Its Development to Man

In the Universe have occurred three vast revolutionary events which radically changed its character. The first was the formation of matter. Thin clouds of hydrogen permeating space to the ends of time condensed under the force of gravity, became dense enough to generate enormous pressures, fused in clusters of four atoms to form helium, with the release of enormous energies, and then used these energies to build larger and larger atoms until all of the elements of the Periodic Table were created.* This self-creation continues today; for the same conditions hold, although with less free hydrogen, and the same forces exist. The Earth is but an infinitesimally small chunk of matter thrown off by a small whirling sun and revolving around it in the more or less stable equilibrium of a slightly imperfect orbit.

The second revolutionary event was the creation of life. To have life there must be water. Fortunately the earth was not a smooth ball. When two hydrogen atoms reacted with one of

---

*Elements with atomic weights of 7-100 (lithium to molybdenum) were formed mainly by the fusion of hydrogen and helium atoms; heavier elements were formed largely by neutron capture by lighter elements.

oxygen, with explosive force, there were enough mountains and deep valleys to collect the water formed into primitive seas, which cooled the earth and helped to equalize the daily fluctuations of temperature from boiling hot by day to freezing cold by night. Thus life had a stable environment in which to begin, the sea.

No one has explained satisfactorily the marvel of the creation of life. Because we know the structures of living things, at least their components, we know what was needed in the beginning: water. Then an element which would bond with itself in a long chain; there are only two, carbon and silicon, and the latter is unsuitable, as we have discussed. A combination of carbon with oxygen, and hydrogen with nitrogen. Oxygen and nitrogen. Carbon and hydrogen. And later, when complexities were needed, phosphorus and sulfur compounds, and the metals potassium and sodium, magnesium and calcium in solution. In the primordial seas were all of these elements, and the atmosphere was probably a mixture of carbon dioxide—soluble in sea water—methane and oxygen. Hydrogen was all combined as water, methane, and ammonia, as it is too light to remain in the lower atmosphere.

In an órderly manner, with enormous wastage, short chains of carbon and hydrogen atoms combined with ammonia and carbon dioxide to form amino acids, and these amino acids, probably catalyzed by the abundant metal, magnesium, or by the trace metals manganese or zinc, combined in long chains to make proteins. There were many varieties, but a few dependent on metals acquired the ability to act as catalysts themselves, or enzymes. When one brought zinc into its large molecule, it was able to join with other molecules and grow. Growth is the first requisite of life, reproduction the second. Whether by chance, cataclysm of Nature or Act of God, simple organisms not unlike viruses were created and reproduced—we know not how—and suddenly the seas were filled with living matter, using those elements which were available in the environment. This poorly told tale makes no mention of how or why, only what.

The primitive ocean contained those elements which we

find today. But it was much more dilute than is modern sea water. At its beginning 3.3 billion years ago, it was too dilute and too hot to support life. As the heavy rains washed soluble elements from the land masses into the seas, from which water was evaporated by the sun only to fall as rain on the land and carry soluble elements into the seas endlessly, as happens today on a smaller scale, the oceans accumulated the elements, particularly sodium and chloride, magnesium, calcium, potassium, sulfates and phosphates. When concentrations were right, plant life began. It is likely that for two-thirds of the earth's history its surface was lifeless. The earth is only four and a half billion years old,* life over two billion years, a fairly short interval in the history of the Universe since the first creation of matter, and two fifths of the time elapsed since the condensation of our little sun from hydrogen.

Table II-1 shows the major elements in sea water now; then they were probably diluted to a fifth or less.† But they were available, and the primitive viral-like and bacteria-like organisms multiplied and developed more and more complex systems and structures according to what seems like an extremely well ordained Plan, with enormous wastage on which they fed, until plants evolved and covered the seas. By this time porphyrins, or ring structures, had been synthesized, and one containing magnesium in its center—chlorophyll—had the marvelous ability to absorb the energy of light and use it to take carbon dioxide from air or water, release the oxygen, and form the chains of carbon atoms with water to make what we call carbohydrates, or sugar and wood—or its primitive analogue. We owe to plants the conversion of a carbon-dioxide-rich atmosphere in which

*This age was measured by the decay of the isotopes of lead, and applies to a solid earth. If the earth was formed around the time the sun condensed, a reasonable guess would make it four and a half billion years. Of course God only knows. The oldest life forms date about two billion years ago, although there are some limestones apparently formed by living things which are 2.7 billion years old.

†Among them was magnesium, which stabilizes ribonucleic (RNA) and deoxyribonucleic acids (DNA), the bases of differentiation and of heredity, and upon which exchanges of energy in phosphates depends. Life is dependent upon the conversions of locked-up energy.

TABLE II-1

Essential bulk elements in sea water and in the earth's crust

Element	Seawater ppm	Earth ppm
Chlorine	19,000	200
Sodium	10,500	28,300
Magnesium	1,350	20,900
Sulfur	885	5,200
Calcium	400	36,300
Potassium	380	25,900
Silicon	4	277,200
Phosphorus	0.07	1,180

ppm = parts per million

animals and insects cannot survive into an oxygen-rich atmosphere in which they can. At the same time, the weather and the seas cooled, for carbon-dioxide in the atmosphere exerts a "green-house effect" on the sun's rays, converting light to heat by slowing it down, just as does a window on a cold sunny day in winter.* By the time this conversion was virtually complete, the prodigious growth of plants slowed down because of the relative scarcity of carbon dioxide in the atmosphere, and the formerly choked seas became ready for animal life. Man in his infinite wisdom is now slowly reversing this primordial conversion which made him possible, by burning vast amounts of carbon and its compounds and discharging carbon dioxide into the air.

Plants used all of the eight bulk or major elements in sea water for their structures and chemical reactions, and had early learned to use one element occurring in trace amounts, zinc, for growth. At some time they also found it easier to survive if they

*Some of the released oxygen was converted into ozone, which formed a layer in the stratosphere 18 to 30 miles up. Ozone absorbs short wave ultraviolet radiation, which is lethal to living cells and can break up simple molecules of water and carbon dioxide. The surface of the earth was probably unpopulated until this layer of ozone was formed.

took advantage of another trace metal, copper, and to depend upon another, boron, for complex functions. They may have used iodine and bromine, for there was much of the latter, although this is doubtful. Later they found that they could increase the efficiency of their living systems with manganese and iron. These trace elements became essential for plants and still are.*

We define a trace element for our own purposes as one occurring in amounts less than 0.01 percent of the human body. While this definition is narrow and restrictive, it fits other living things remarkably well, and attests to a certain uniformity in Nature. Elements making up the bulk of a plant or animal make up the bulk of living, growing matter. Elements used in traces in man are used in traces by plants and animals. So we will continue with this definition until a better one comes along.

Let us jump this narrative a billion years after the seas were choked with plant life and look at modern sea water, enriched by all the soluble metals of the earth's crust, from the aspect of the trace elements. The trace elements can be divided into two main groups, the ones which take part in necessary biochemical reactions, or the essential trace elements, and the ones which don't. Those which don't can be subdivided into elements which do no harm to living things—non-toxic—at usual environmental exposures and those which do harm living things. Obviously no element encountered in sea water at usual concentrations is toxic, or there would be no life today. In table II-2 are shown the essential trace elements used by marine plants and animals, and their concentrations in sea water, in parts per million (ppm), or micrograms per gram or milligrams per kilogram or grams per metric ton. (The metric system has helped scientific measurements enormously, and we should abandon the obsolete English system of weights and measures.)

Whereas no element in the concentrations found in sea water is toxic to living things, all elements are toxic in large

*All plant life, including bacteria, fungi, algae and higher plants, require copper, iron, magnesium, manganese and zinc. Green plants and algae need boron, and some require cobalt and molybdenum. A few bacteria, fungi and algae need vanadium.

TABLE II-2

Essential trace elements in sea water and in the earth's crust.

Element	Sea Water ppb	Earth's Crust ppm	Used by marine organisms	Used by land animals
Strontium	8000	450	Vertebrates	Vertebrates
Boron	4600	16	Plants	Plants
Fluorine	1300	700	Vertebrates	Vertebrates
Iodine	50	0.3	Plants? Fish?	Mammals
Zinc	15	65	All life	All life
Molybdenum	14	1	Vertebrates	Bacteria, Vertebrates
Copper	10	45	All life, arthropods, molluscs	All life
Vanadium	5	110	Ascidians (sea squirts)	Bacteria, Vertebrates
Selenium	4	0.09	?	Birds? Mammals
Iron	3.4	50,000	All life	All life
Chromium	2	200	Vertebrates?	Vertebrates? mammals
Manganese	1	1000	All life	All life
Cobalt	0.1	23	Vertebrates	Mammals

Note: The most concentrated elements are the least toxic, strontium, boron, fluorine and iodine. ppb = parts per billion. ppm = parts per million.

enough amounts, with rare exceptions, because of being insoluble or completely inert. Gabriel Bertrand, a French scientist and the father of the study of trace elements, proposed a rule which we will call Bertrand's Law of optimal nutritive concentration. It states that in the absence of an essential element a plant cannot live. It thrives on adequate amounts, but an excess is toxic. He worked this out for manganese, but the principle applies not only to plants but to all animal life. Figure II-1 shows this Law graphically. Eugene D. Weinberg has discovered an extension of the Law which we will call Weinberg's Principle. He showed with certain bacteria and manganese that amounts adequate for growth were not necessarily adequate for optimal function—in this case the production of an antibiotic—and this Principle bears very important applications for the nutrition and health of all living things, especially us.

But as I said, Bertrand's Law and Weinberg's Principle are not applicable to sea water, for nothing is present in suboptimal or toxic amounts, except in places where Man has discharged effluents into rivers which foul estuaries with toxic amounts of essential and abnormal elements. (See Figure II-2.) (The delectable Maine lobster is growing scarce and his price is inflating owing partly to such practices, for he cannot survive in polluted water.)

A glance at Table II-2 will show the idiocy of the dogma pounded at our ears by anti-fluoridationists over and over until it has become the Big Lie: Fluoride added to drinking water at 1 ppm is toxic; sodium fluoride is the "unnatural form" of fluoride; calcium fluoride is the "natural" form. The truth is that almost all of the fluoride in the sea is in the form of sodium fluoride; that calcium fluoride is insoluble in water, that there are 1.846 billion tons of sodium fluoride in the oceans, that the seas are teeming with life of all kinds exposed constantly to 1.3 ppm fluoride, that life began in the presence of fluoride, of which the earth's crust contains 700 ppm. If fluoride were as toxic as these faddists say it is, I would not be here writing this book, for I never would have evolved. Fluorine

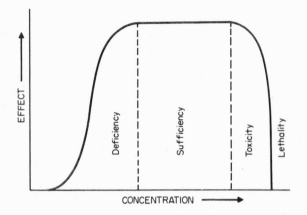

Figure II-1: *Schematic illustration of Bertrand's Law of optimal nutritive concentration of an essential trace element. When there is no element present there is no growth or function. The first part of the curve indicates a state of deficiency which becomes less as the concentration increases. The plateau indicates a state of sufficiency of the element when function is at its highest level. The width of this plateau varies with different elements and living things, but in mammals is wide, as the animal can get rid of excesses which it does not need. When high concentrations exceed the ability of the animal to repel or excrete excesses, the element becomes toxic, and is lethal at the point where the curve descends perpendicularly. This curve applies to all living things and may be quite narrow in the case of plants and especially marine organisms. It is important to note that growth is not necessarily a sign of optimal function when the element is somewhat deficient. For example, a bacteria may grow well when an element is partially deficient, but may not produce compounds which it ordinarily would produce in the presence of sufficient element.*

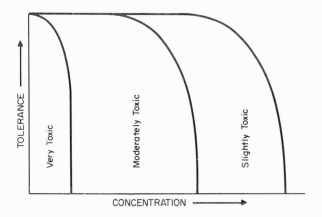

Figure II-2: *Schematic illustration of effects of toxic trace elements which have no biological function. A certain concentration is tolerable but slight excesses may show increasing toxicity and eventual lethality. The first curve on the left, for example, can represent cadmium or mercury for which there is a definite but small tolerance. The middle curve could represent lead for which there is more tolerance. The curve on the right can represent antimony or tin, for which tolerance in the mammal may only be evident by a shortening of life span. For toxic elements there are differing thresholds at which effects appear; this phenomenon is common to all living things. In actual fact, all soluble elements can be toxic in large enough amounts.*

in the form of its most abundant salt, sodium fluoride, is an element which promotes the strength of bones and teeth, and marine vertebrates benefit from its presence even if people living in low fluoride areas who don't fluoridate their drinking water don't.

Let us go back in time from the Twentieth Century to five

million centuries ago, or thereabouts. At that time, primitive pre-chordates,* invertebrate worm-like animals with spinal cords but few if any brains were mobile, swimming in vast numbers in the teeming sea. Their needs were simple. Their skins were permeable to water, salts, and nitrogen-carrying waste products; they merely swam away from their excreta. They had an intestine of sorts and a way of absorbing oxygen through their skins. They had primitive blood. Perhaps because of a population explosion, some of the more adventurous of them began to invade brackish water and later to swim up the swiftly flowing Cambrian rivers.

To do this, they needed several important additions. First, a skin impermeable to water and salts, for life began in salinity and cannot free itself therefrom. Fresh water would dilute their blood and cells and they would die. Their bodies, and each cell in their bodies, were surrounded by a membrane which was, and is, permeable to water, most salts and small organic molecules, but not to larger ones such as proteins. The "osmotic pressure" or the pressure exerted by charged particles in the fluids outside the cells always equals the pressure inside the cells—this important physiochemical relationship is known as the Donnan Equilibrium. It applied equally to these primitive creatures as it does to mammals. If the outside saline solution is diluted, the cells and the body swell with water, as you know if you have seen fresh oysters "floated" in fresh water, a practice which enhances their beauty but spoils their taste. If the saline fluid outside the cells is concentrated, the cells and the body shrink, or dry up. Life depends on this equilibrium being constant, for cells cannot function when dry or over wet.

Second, they needed a gill for extracting oxygen from fresh water and repelling the water. Third, a kidney for excreting the excess water they drank, and for extracting from it the necessary elements they formerly absorbed through their permeable skins. Fourth, a bony skeleton so that they could

*Plants feed on matter and animals feed on plants. The first fossil Protozoa, an animal of sorts, lived about 800 million years ago in the midst of an abundance of food. It took only 300 million years for organized animals to evolve.

swim strongly against the current. Fifth, a brain so that they could know where to swim. Evolution of vertebrates which could carry with them a skinful of primitive ocean into an adverse environment to which they could adapt was a tremendous leap forward, for it made possible the eventual invasion of the land.

These creatures, our ancestors, which we now call fish, remained in fresh water for a geological period or so. Then, for some unknown reason, perhaps the coming of an ice age, they returned to the sea. By this time, the sea had become much saltier, and the primitive oceanic environment with which their cells were bathed within their impermeable skins was much more dilute than the surrounding sea. All of the marine invertebrates—animals without an impermeable skin, a backbone and a brain—had adapted their cells to the increasing salinity of the ocean, but these fish were made up of cells evolutionally accustomed to the fixed saline environment at the time they left it; and they could not survive in more salty water. So they developed a gill which would reject sodium chloride at its surface and they restyled their kidneys to excrete any excess salt they might swallow. Thus, they preserved the salinity of the primitive ocean within their bodies, and this degree of salinity is universal in all vertebrates today, with minor variations.*

The shark and other similar fish were apparently slow movers, never developing a bony skeleton, for example—they use cartilage—and they remained in fresh water long after other fish had left. In doing so, they became evolutionarily fixed, and when they finally returned to the saltier sea, they had lost their ability to change their gills to reject salt. They had only one other choice for survival. They could increase the total concentration of soluble substances in their blood and extracellular fluids to equal the total concentration of salt in the sea. This they did, by developing a kidney which would retain urea in their bodies. Sharks today are uremic, and this

*We can infer that the sea was about a third as salty in Cambrian times as today. Mammalian serum contains about 3280 ppm of sodium, sea water contains 10,500 ppm.

end product of nitrogen (protein) metabolism, which is made up of ammonia molecules, is present in shark blood to the extent of 2 percent or a hundred times the amount found in mammals and other fish. Urea splits easily into ammonia, which is why public toilets, babies diapers and warmed shark meat smell of ammonia.

When these drastic changes were evolving, the trace elements were not neglected. In fresh water sources were available in water, in animal food and in plants, and mechanisms for absorbing and retaining needed amounts and excreting excesses taken in became highly developed in gill, gut and kidney. Those mechanisms exist today. The intestine was the principal route of excretion of excess magnesium, chromium, strontium, manganese, iron, copper and zinc; the kidney the main route for sodium, potassium, calcium, borate, molybdenum, fluoride, phosphate, sulfate, chloride, selenate and iodide.

The most ancient method for excretion of trace metals was developed in primitive shellfish long before vertebrates evolved. It is still used in mammals. With the evolution of the primitive gut, and the more sophisticated but primitive oyster, clam, mussel and the like, the first important organ for digesting food within the body came into being as a budding from the gut. This was the hepatopancreas, a combination of liver and pancreas which had a duct discharging into the intestine. (Today, the liver and pancreas have separate ducts, joining into the "common bile duct.") In order to excrete potentially toxic metals absorbed into the body, the hepatopancreas was used as an outflow tract, the metals being bound to what later became bile. In mammals, today, manganese is excreted via bile and pancreatic juice, and bile contains chromium, copper, zinc and most metals not necessary for the body.

I do not want to give the impression that the evolution of vertebrates proceeded regularly according to a previously conceived plan. Countless numbers of trials and errors took place, using the elements available; some succeeding to continue evolving, some partly succeeding only to end in evolutionary stasis, many failing. All living matter had the evolutionary stimulus, and probably has today, an urge to further complexity, a restless force to adaptation and readaptation in response to environmental changes. Success depended upon the right choices made at the right time.

If I may personify these protoplasmic aggregations, I can illustrate this matter with two wrong choices leading to evolutionary confinement or stasis. A wrong choice led to a blind end of the road. Most molluscs chose copper for their oxidation-reduction systems and for carrying oxygen in their blood much more than half a billion years ago (exact date not recorded). Copper has considerable disadvantages, for it has only half the oxygen carrying capacity of iron. Copper was also chosen by crustaceans. Copper protein complexes are contained in the blood of molluscs, arthropods and trilobites, and was widely used by many animals in the Paleozoic era. Worms chose iron to carry their blood oxygen; being twice as efficient, it allowed the evolution of the vertebrates ending in that prime vertebrate of all, Man. Molluscs which chose iron, as some did, can live out of water—snails for example; younger species of molluscs more recently developed must have learned from the experience of their ancestors, for they also chose iron.

The first chordate fish, our ancestor, was amphioxus. This primitive vertebrate had enormous potentialities. It had a notochord, a spinal column with a bulge on the front end from which developed the most complex organ on Earth, the human brain. Amphioxus was on the right track, for he chose iron for oxygen systems. One of his sisters, however, was less conservative, more curious, and more adventurous. She chose vanadium to carry her blood oxygen. Vanadium was the wrong metal, for instead of ending up as a human being she ended up as a sea squirt. These ascidians have green blood cells full of vanadium, up to 4 percent, bound in complexes like bile pigment. Vanadium is adequate for sea squirts but not for anything more complicated. The sad part of the story is that their larvae have notochords, but as they grow, they lose their nervous systems, degenerating into sea squirts with only a single nerve ganglion which lacks the first faint capacity for thought. Thus the choice of vanadium induced an evolutionary regression, an experiment which came to a blind end because of the chemical limits of this trace metal as a versatile catalyst for living things.

From these accounts, we can see that when vertebrates decided to invade the oceans' shores, those of their chemical systems depending on bulk and trace elements were evolutionarily fixed, and there was no need to develop them much further. The proper amounts of trace elements were present in sea water, not too much and not too little. There was a problem with excess of bulk elements which had been solved by gill, gut and kidney, and a problem with excesses of trace metals, early solved by the hepatopancreas; and plenty of food was at hand. There seemed to be no earthly reason why they should not be complacent and happy in the environment where they had evolved. But they were not. They chose or were forced to invade the land.

CHAPTER III

Trace Elements
in Mammals

It took about 250 million years for fish to evolve into amphibians and amphibians to evolve into reptiles. All of this time was spent on the littoral or in estuaries and swamps, in salt or fresh water. The primitive amphibian could lay its eggs in water, where they developed into fish, grew legs in place of fins, invaded the land for very short distances until their skins were dry, and returned always to water, much as does the modern frog, which mirrors this process in a few months or a year. By this time, plants had adapted to dry land and they and insects had taken over the earth.* In respect to the elements, amphibians had few problems, for there was always the water available—the sea, with its plentiful supply of bulk and trace elements and the rivers from which fish had learned to extract elements two hundred million years or so previously.

At some Jurassic time when the first terrestrial reptilian vertebrate, our ancestor, left the littoral and walked or slithered on land, the problems he met were enormous. He carried with him a skinful of the primitive Cambrian Ocean with the

*Insects require copper, iron, managanese, zinc and probably cobalt and molybdenum. Higher plant-eating insects have low levels of sodium in their blood and high levels of magnesium; they may have evolved on land where it is difficult to obtain much sodium but easy to find potassium and magnesium in plants. Their muscles contract and relax in an unique environment, unlike all other animal life. Primitive insects depend on sodium.

composition and concentration necessarily fixed, but no longer could he return to the sea to replenish it. From an environment of evenly distributed plenitude he entered an environment of uneven areas of superabundance and deficiency. To obtain an adequate supply of an essential element, he often had to travel long distances; therefore he needed to develop mechanisms to conserve what he had in the face of deficiencies. The first requirement was a way of conserving water to prevent death from desiccation. The second was a way of conserving sodium, potassium, magnesium, calcium, sulfur, phosphorus.

The new environment, the earth's crust, contained as it does now a superabundance of sodium, calcium, magnesium, potassium, silicon, phosphorus, sulfur and a dearth of chlorine (see Table II-1). There was more fluorine than chlorine. There was a superabundance of iron, manganese, chromium, vanadium, cobalt, zinc, copper, adequate amounts of boron and a relative dearth of molybdenum and iodine (See Table II-2). There was also a superabundance of aluminum, titanium, rubidium and barium. There were intermittent supplies of water.

In respect to certain trace elements distribution was uneven. Zinc, copper, iron and manganese were plentiful wherever there was vegetation; cobalt, molybdenum, fluorine and boron were not distributed uniformly. Therefore, in order to survive, this land-based animal had to learn not only to conserve essential elements, but to reject excesses which could be toxic. The magnitude of these differences is shown in Table III-1 and II-2, and for the bulk elements, the daily need or intake (Table III-1).

In the face of these differences, exquisitely sensitive homeostatic mechanisms were developed in gut and kidney, which conserved deficiencies and rejected excesses. Homeostasis means the same state, or a balanced equilibrium, which will be further explained in the next chapter. These mechanisms were necessary for survival, allowing the evolutionary process to proceed to reptiles and birds, mammals and finally to man. It was in man that the third great revolution occurred, the

TABLE III-1

Major elements in the body of man and in sea water.

Element	Seawater ppm	Amount in body grams	Total body ppm	Daily need, g	General Location
Organic					
Oxygen	875,000	43,000	610,000	2550	All tissues
Carbon	28	16,000	230,000	270	All tissues
Hydrogen	108,000	7,000	100,000	330	All tissues
Nitrogen	0.5	1,800	26,000	16	All tissues, protein
Inorganic					
Calcium	400	1,000	14,000	1.1	Bone, outside cells
Phosphorus	0.07	780	11,000	1.4	All cells, inside
Sulfur	885	140	2,000	0.85	All cells, inside
Potassium	380	140	2,000	3.3	All cells, inside
Sodium	10,500	100	1,400	4.4(0.2)	Extracellular fluids
Chlorine	19,000	95	1,200	5.1(0.4)	Extracellular fluids
Magnesium	1,350	19	270	0.31	All cells, bone, inside
Silicon	4	18	260	0.003	Skin, lungs

TABLE III-2

Trace elements in seawater and in man, ancient and modern.

Element	Sea Water ppb	Primitive Man ppm	Modern Man ppm	Principal Cause of Difference
Essential				
Iron	3.4	60	60	
Zinc	15	33	33	
Rubidium	120	4.6	4.6	
Strontium	8000	4.6	4.6	
Fluorine	1300	37	37	
Copper	10	1.0	1.2	Copper pipes
Boron	4600	0.3	0.7	Vegetables and fruits
Bromine	65,000	1.0	2.9	Bromides? Fuels
Iodine	50	0.1–0.5	0.2	Salt iodized
Barium	6	0.3	0.3	
Manganese	1	0.4	0.2	Refined foods?
Selenium	4	0.2	0.2	
Chromium	2	0.6	0.09	Refined sugars and grains
Molybdenum	14	0.1	0.1	
Arsenic	3	0.05	0.1	Additives, weed killers
Cobalt	0.1	0.03	0.03	
Vanadium	5	0.1	0.3	Petroleum

Non-Essential			
Zirconium	0.02	6.0	6.7	
Lead	4	0.01	1.7	Motor vehicle exhausts
Niobium	0.01	1.7	1.7	
Aluminum	1200	0.4	0.9	Food additives
Cadmium	0.03	0.001	0.7	Refined grains, water pipes
Tellurium	?	0.001	0.4	Metallurgy
Titanium	5	0.4	0.4	
Tin	3	< 0.001	0.2	Tin cans
Nickel	3	0.1	0.1	
Gold	0.004	< 0.001	0.1	Ornaments
Lithium	100	0.04	0.04	
Antimony	0.2	< 0.001	0.04	Enamels
Bismuth	0.02	< 0.001	0.03	Drugs
Mercury	0.03	< 0.001	0.19	Fungicides
Silver	0.15	< 0.001	0.03	Eating utensils
Cesium	2	0.02	0.02	
Uranium	3.3	0.01	0.01	
Beryllium	?	< 0.001	0.001	Fumes and smokes
Radium	0.3×10^{-10}	4×10^{-10}	4×10^{-10}	

creation of conscious thought, which began his further evolution towards what he will become in the next half million years—unless he is supremely foolish.

To this point we have concerned ourselves with the elements essential for living things. But there are many more, as we have seen in Chapter I, some of which may be beneficial, some inert and some toxic. We will consider those elements found in the body of man in varying amounts, all of which are in sea water, to all of which life has been exposed from the beginning of time, and all of which, toxic or inert, have occurred in sea water and on land in concentrations which were non-toxic. In table III-2 are shown the concentrations of the essential trace elements in primitive man as compared to those in sea water. As we have no analyses on primitive man, and as modern man living in truly primitive environments is fast disappearing from this earth, so widely has civilization spread, that we were forced to do a bit of guesswork on this section of the table. We have assumed that the requirements of primitive man for essential trace elements were similar to those of modern man, and we have attempted to evaluate modern exposures to elements as by-products of modern civilization. Therefore, we can accept these figures as educated guesses.

In the next column of the table are the concentrations of elements found by analysis in modern human bodies. Our present civilization began some 4500 years ago when metals were first fashioned and used as aids to living. Since then, mankind has dug from mines and deposits vast quantities of elements, metallic and non-metallic, some of which he has spread over the face of the earth to contaminate his food, his water and his air. Although all of them are found in sea water, exposures were usually minor both on land and sea when mining first began, and there is no reason to believe that protective mechanisms for repelling these elements and adapting man to them developed in such a short time as 4500 years.

Metals known to the ancients were believed to possess magical or spiritual powers and the stars in the heavens were linked with them. According to no less an authority than the

TABLE III-3

Some trace elements in wild animals and man

Element	Modern Man ppm	Wild Animals ppm
Iron	60	60
Zinc	33	22-30
Copper	1.2	1-9
Manganese	0.2	0.7-2.4
Chromium	0.09	0.11-0.48
Arsenic	0.1	0.4-1.1
Cobalt	0.03	0.05-0.2
Vanadium	0.3	0-2.0
Lead	1.7	0.25-0.8
Niobium	1.7	3.8-4.7
Cadmium	0.7	0.1-0.3
Tellurium	0.4	0-0.4
Titanium	0.4	0.3-3.0
Tin	0.2	0-2.2
Nickel	0.1	0-0.2

canon's yeoman of Chaucer's Canterbury Tales, the Four Spirits are mercury, arsenic, sal ammoniac and brimstone, while the Seven Bodies of the heavens are gold (the Sun), silver (the Moon), iron (Mars), mercury (Mercury), lead (Saturn), tin (Jupiter) and copper (Venus). You will note that two of these are essential for life, two are inert and three toxic. The ancient symbol for copper is the crux ansata, ♀, or the ankh, which also signifies femaleness, Venus and eternal life. The priests in Verdi's opera *Aida* point inverted ankhs to declare death to Radames.

In the fourth column I have attempted to list the major cause of the differences in concentration of each element in modern as compared to primitive man, who we must assume was healthy. Among the essential or possibly essential elements there are few differences, but these may be important. In modern man there is a little more copper, from increased exposures, mainly in his liver. There may be more boron, because of the prevalent use of vegetables and fruits. In people living away from the sea there is more iodine, because of the addition of iodine to salt. There is a bit more arsenic due to insecticides, pesticides and weed killers of arsenic. There is more vanadium, from the burning of petroleum. There is less, much less, and probably insufficient amounts of chromium, owing to the use of refined sugars and grains (see Chapter VI).

Among the non-essential elements, more marked increases in modern man are evident. There is a large increase in lead, today from motor vehicle exhausts, formerly from water pipes and food containers. There may be a bit more aluminum from food additives, much more cadmium from refined grains and water pipes, much more tellurium from metal ores, much more tin from tin cans, more gold from jewelry, more antimony from enamels, more bismuth from stomach-soothing drugs made necessary by the tensions of modern life, more mercury from fungicides, more silver from eating utensils, more beryllium from smokes, more germanium from coal and more of almost everything man uses today.

In order to estimate roughly the increments or losses of trace metals in modern man, we can compare his concentrations with those in the tissues of wild animals living in a remote Vermont forest (Table III-3). The animals included mice, squirrels, woodchuck, deer, beaver, fox, raccoons. Of the essential metals, wild animals had more chromium, copper, cobalt, and manganese than human beings and about the same amounts of iron and zinc. They had much less lead, cadmium, tellurium, somewhat less nickel, and probably more vanadium, niobium, titanium, arsenic and tin. One can infer from this table that the industrial civilization and what it has done to foods has resulted in decreased exposures or losses of chromium, copper, cobalt and manganese, and increased exposures of lead, cadmium and tellurium. This comparison assumes that the animals were healthy and that their homeostatic mechanisms were active.

Do these increased or decreased concentrations do us any harm? That question is what the rest of this book is about.

CHAPTER IV

The Essential Trace Elements

Elements which are necessary for biological function are known as essential, that is, without them, life does not exist. In terms of optimal function of living things, however, we must broaden this definition to include elements necessary for special purposes which provide perfect health but without which life can still exist at an unhealthy level. We do not know all of the elements of this second type, nor in fact are we certain of all of the first. Essential elements are being discovered at a rate of two each decade, and 40 years ago there were only two trace elements recognized; now we know of nine or ten. Just as human beings can exist when they are partly deficient in one or more vitamins, deficiency disease not necessarily being fatal, so can they exist in a state of disease caused by deficiency of an essential element.

By the most rigid definition, no trace element essential for life has been proven to be essential for human beings. In other words, no one has deprived a group of people of an element to the point that deficiency resulted in death. Likewise, human experiments depriving the body of essential vitamins and trace elements are seldom done, for the consequences could be hazardous if not disastrous. Therefore, minimal human requirements of many trace elements and many vitamins are not known. Committees of nutritionists who decide minimal requirements, however, seem to stick subconsciously to the rigid definition of essentiality of some vitamins and trace elements about which less is known than about others, and refuse to state that they are necessary for optimal health. Pyridoxine, or vitamin B_6, is a good example. Because human requirements are not exactly known, although rat and rabbit and guinea pig and chicken and turkey and monkey requirements are known, B_6 has been left in a limbo of vagueness until recently, in spite of the fact that it is depleted in milk by irradiation and pasteurization, partly destroyed by cooking, removed in the refining of flour by two-thirds or more, and is not put back into foods from which it is removed. Committees seem to ignore the fact that B_6 is necessary for the smooth integrity of the walls of arteries and the function of nerves. There are a number of examples of similar bureaucratic inconsistencies and nonsenses.

Human requirements for most of the bulk elements are known, for disorders have resulted from depletion and have been cured by restoration. Amounts in the earth's crust and in the "Standard" human body and the daily requirements of man are shown in Table IV-1. There is plenty of these elements in sea water to allow for adequate amounts in the evolution of vertebrates and especially mammals. Some deficiencies are from natural causes; excess loss of sodium in sweat, for example, results in muscular cramps. Poor and abnormal diets may be low in calcium and possibly magnesium, but any well rounded diet contains more than enough of the bulk elements to supply needs. The ranges of each element which can be tolerated and still maintain a balance in the body are relatively enormous, and only in diseased states are secondary disturbances found, owing to disorders of homeostatic mechanisms.

Homeostasis (same state) is a condition of balance. When applied to the elements, it means that intake and outgo are equal. The body has marvelously exact mechanisms for maintaining homeostasis of its inorganic elemental composition. Unfortunately it has relatively poor mechanisms for maintaining the organic elemental content of the carbon, hydrogen and oxygen which is built up and deposited as fat. If we eat more than we burn, we gain weight. If we eat less than we burn, we lose weight. There are no magic formulae for weight gain or loss; it is as simple as that. But homeostatic mechanisms are highly efficient for magnesium, sodium, potassium and calcium, and probably for sulfur, phosphorus and chlorine.

These mechanisms are located in the intestinal tract and in the kidney. The small intestine absorbs, rejects, and excretes in its juices; the large intestine reabsorbs what the body needs as it absorbs the fluid contents of the small intestine. The kidney both excretes excesses in blood and retains when blood contents are low. Elements dissolved in blood plasma can be excreted in sweat, for sweat glands are like minute primitive kidneys. If we swallow more sodium than the body needs, the large intestine absorbs less and the kidney excretes the excess in blood. The range of tolerable limits for sodium is 0.2 gram to 18 grams in temperate climates, or a difference of 90 times. Potassium is easily absorbed from the intestine, but there is little in blood. It is stored within the cells of the body where it is held largely by action of the adrenal glands; excesses in blood are excreted in urine. Excesses of magnesium or retention by the kidney when intakes are low is so efficient that less than 4 percent of the normal intake can maintain balance. Calcium behaves somewhat like sodium. These four elements are under glandular control.

Trace metals are also under homeostatic control because accumulation of excesses are toxic and deficiencies unhealthy. Iron, copper and manganese controls are known; controls for chromium, zinc and perhaps cobalt undoubtedly exist, although they have not been described. No controls for fluorine, iodine, molybdenum, or selenium can be postulated at this time, the body depending upon outside sources and becoming deficient when the environment is deficient. But for most of the metals, homeostatic mechanisms are marvelously efficient.

One very ancient method of excretion used by oysters and man and probably every living animal in between is by way of the bile and pancreatic juice, which empty the digestive juices

TABLE IV-1

Essential bulk and trace elements on the earth's crust and in man.

Element	Crust ppm	Man ppm	Daily requirements in food and water g	Homeostasis in man
Bulk or major				
Calcium	36,300	14,000	1.1	+
Sodium	28,300	1,600	4.4(0.2)	+
Silicon	27,720	260	0.003	?
Potassium	25,900	2,000	3.3	+
Magnesium	20,900	290	0.31	+
Sulfur	5,200	2,300	0.85	+
Phosphorus	1,180	12,000	1.4	+
Chlorine	200	1,400	5.1(0.4)	+

Element	Crust ppm	Man ppm	Daily requirements in food and water g	Homeostasis in man
Trace				
Iron	50,000	60	0.013	+
Manganese	1,000	0.2	0.003	+
Fluorine	700	37	0.003	0
Chromium	200	0.2	0.0005	?
Zinc	65	33	0.013	+
Copper	45	1.0	0.005	+
Cobalt	23	0.02	0.0003	+
Molybdenum	1	0.1	0.0002	?
Iodine	0.3	0.2	0.0001	0
Selenium	0.09	0.2	0.00001	?
Vanadium	110	0.3	?	?
Possibly essential trace				
Nickel	80	0.1	?	?
Arsenic	2	0.1	?	0

of the liver and pancreas into the small intestine. We have already mentioned this method. Because life began and evolved in sea water, the major constituents sodium, potassium, magnesium, calcium, strontium, boron, bromine, fluorine, and aluminum could not be toxic at the concentrations existing then and now—or life would have been extinguished. As more and more complex organisms evolved and as specialized structures developed, such as the primitive liver and gut, a tendency to accumulate metals encountered in trace amounts appeared, and some method for their excretion from the liver became necessary. Thus was evolved the first primitive control to excrete excesses into the gut. Such controls were unneeded for the major components of sea water until marine animals left the sea to invade fresh water, and later, the land. This primitive mechanism, more highly developed, has persisted today in mammals, and is used to rid the liver of excesses of copper, zinc, and manganese, and some toxic trace metals, by excretion in bile. Because practically all of the blood supplying the stomach and the intestine, into which digestive products and elements are absorbed, returns to the liver, many of the trace elements are partly screened out by the liver, returning to the intestine in bile.

Homeostatic mechanisms can fail in two directions under the following conditions:

1. When the amount of an element taken into the body daily is less than the amount excreted obligatorily in feces and urine. There is always enough leeway, through the use of body stores of the element, to allow for a considerable period of time to elapse on what is called a "negative balance" and still maintain health, but eventually a state of deficiency will develop. If one spends a little more money than one makes each month eventually one's bank account will be empty. The body, however, cannot "borrow" trace elements; it has no credit but its stores. Certain diseases can increase the excretion of trace elements, making the need for larger quantities real. Chronic disorders of the kidney can interfere with element-saving mechanisms for those largely excreted or saved by the kidney, thus increasing obligatory losses. Chronic diarrhea and poor intestinal absorption can increase the need. Alcohol in

continuous amounts can mobilize zinc, for example, from the liver and magnesium from other tissues, increasing excretion and outgo.

2. An amount in excess of the intestine's ability to reject or the kidney's ability to excrete will lead to accumulation of an element in the body and eventual toxicity to cells or to an organ. The nature of the toxicity depends upon the affinity, or lack of it, for special cells or special organs in which the element accumulates. The storage banks become overloaded with the element, and sometimes it takes as long to get rid of the excesses as it did to accumulate them, i.e., many years. For example, some South African natives eat relatively enormous amounts of iron absorbed into beer and food from iron pots and storage jars, and accumulate enough iron to cause toxicity. The homeostatic mechanism for retention of iron is extremely efficient, or to put it another way, that for excretion of iron is very poor, so the tissues hold iron avidly, and removal is most difficult. Therefore, excesses can be worse for health than deficiencies, which can be made up.

3. Hereditary diseases in which the transporting mechanisms for a trace metal are inefficient, especially the intestinal rejection of the metal, allow accumulation and serious disease under normal exposures. Two are known, for iron and for copper. The gene carrying the mechanism is injured in some way. Hemochromatosis is the outcome of the disorder for iron which accumulates slowly throughout life, damaging the pancreas (diabetes results), the liver (cirrhosis results) and the skin (bronzing results). The rate at which the disease develops depends on the degree of damage to the mechanism for absorption of iron; the disease usually appears clinically in the forties or fifties, with death inevitable. Because death occurs after the age of procreation, this hereditary disease is likely to increase in the population.

Wilson's disease is the result of accumulation of copper at normal exposures. It is stored in the liver and in special parts of the brain, producing cirrhosis of the liver and a most distressing kind of neurological disorder characterized by constant flapping movements. Fortunately for the race, but unfortunately for the subject, death or disability usually occurs before the age of procreation, and the disease is not increasing much in the population. Because the copper transport mechanism is carried in a recessive gene, and one normal gene is enough to provide good copper homeostasis, it is only when two people each carrying the defective gene mate, that the children have the disease.

No genetic defects in the transport of manganese, chromium, cobalt, molybdenum or zinc have been discovered, although some of them may exist and account for diseases of obscure cause. Because zinc has such little toxicity, and is so vital to life, it is doubtful that hereditary zincism exists. Cobalt is so readily excreted by the kidney that we find it hard to imagine an accumulative disease of this element. So is molybdenum; although an excess might conceivably be remotely concerned with gout, this idea is highly speculative. Hereditary manganism is a distinct possibility; manganese miners absorbing much metal in their lungs accumulate large amounts and show the typical signs of Parkinson's disease, which is fairly common in old people and can have a hereditary tendency. We have no idea what accumulation of chromium would do.

Not all of the functions of the trace elements are known by any means. We can review what is known about their principal functions briefly.

Iron is necessary for the carrying and exchange of oxygen in the blood of mammals, and for many systems involving oxidation and its converse, reduction. There is plenty of iron in sea water, the earth's crust, plants and many animals. It is an ideal metal for this purpose, having two states of oxidation which can switch back and forth with little energy needed (it rusts in air easily). Most of the body's iron is outside of the cells, but inside special blood cells—the red cells. It is a vital element in that without it one dies.

Zinc appears to be necessary for the building of protein, which makes up most of the solid matter of the cells. It is a constituent of the enzyme which releases carbon dioxide from bicarbonate in the blood, of the one which begins the oxidation of alcohol and of other natural substances and of the splitting of portions of proteins. Almost all of the zinc is inside the cells, where it is more abundant than any other trace element. Human and animal sperm contain large amounts of zinc, up to 0.2

percent, and the visual part of the eye contains up to 4 percent. It occurs in high amounts in the eyes of all species examined, including fish. It must be there for a purpose but no one knows what. It is a vital element.

Copper is an excellent catalyst for oxidation-reduction systems, showing great versatility for an impressive variety of reactions, including the formation of water from oxygen and hydrogen at body temperatures; this reaction would be explosive without copper. It is an unique agent in biological systems, all living things require it, and it is vital.

Manganese is required by all living organisms, but its actions are not too well known. It takes part in a number of enzymatic reactions. Deficiency in animals and birds affects bone, reproduction and brain, with abnormalities of bony growth, stillbirths, early deaths and sterility of the mother, and convulsions.

Chromium is needed for sugar and fat metabolism in which insulin takes part. It is necessary for maintenance of normal cholesterol metabolism and normal sugar metabolism. It increases growth. Although animals may live with severe chromium deficiency, they would be most unhealthy and would not grow much. Therefore, it may or may not be vital.

Cobalt's principal function is as a necessary constituent of vitamin B_{12}, which is manufactured by bacteria and needed by mammals. Minimal daily requirement of cobalt in B_{12} is 0.043 micrograms per day, the smallest amount of any substance known to make the difference between health and disease. B_{12} is required for the formation of red blood cells. Other functions of cobalt are suspected but unproven. It is vital in B_{12}.

Molybdenum is a constituent of an enzyme used in the last stage of the metabolism of purines to uric acid. Purines become energy filled structures essential to many exchanges of energy, losses and gains. Molybdenum takes part in the metabolism of toxic aldehydes. It interacts in some way with copper. It may not be vital.

Iodine has only one role in the body, to complete the formation of thyroxin, the thyroid hormone, which contains four atoms of iodine per molecule. Deficiency leads to overactivity of the thyroid, which cannot make the hormone and becomes enlarged into a goitre. It is not a vital element, but its absence causes ill-health.

Fluorine is necessary for the formation of strong, hard bones and for teeth which can resist decay. Deficiencies which are geographical and widespread contribute to decayed teeth, with a consequent demand for more dentists, and probably too brittle bones in the elderly, resulting in fractured hips and much disability. It is not a vital element, but its absence can result in ill-health.

Selenium and vitamin E interact, at least one of them being necessary to prevent muscular dystrophy in experimental animals and growth in chickens. It is essential for mammals even when E is absent. It is not vital, as far as is known.

A number of scientists have suspected that elemental variations in soil might be reflected in food crops and that deficiencies in man might result from foods grown on soils with very low amounts of one or more trace elements. Contrarily, excesses in man from foods grown on soils rich in trace elements might be toxic. Let us examine this idea as it concerns the essential trace elements only, not the toxic ones which are natural or man-distributed contaminants.

It is evident that a trace element necessary for the growth, flowering and seeding of plants must be present in the soils on which food crops are grown, or the plants would not grow and produce seed, roots and fruit. Plants will not live without magnesium, for this bulk metal is an integral part of chlorophyll, the green pigment of leaves (similar to hemoglobin in blood) on which photosynthesis depends. Nor can animals exist without magnesium. Therefore, we are assured an adequate supply of this vital element as long as we eat animal and vegetable products, unless the magnesium has been removed by refining (as it is in sugar and pure fat). The same statement applies to zinc and copper, which are essential for all plants and all animals—all living things.

Manganese is also essential for all living things. Natural deficiency, however, has been described in a number of food plants, which grow poorly and show chlorosis, or bleaching of the leaves. Soils are not deficient in manganese, but when they are alkaline the metal is oxidized to an insoluble form and becomes unavailable. The addition of plenty of organic matter, such as wood shavings, peat moss, manure, compost and the like acidifies the soil, and through the action of soil bacteria, which require manganese for growth, makes manganese available. Deep rooted trees extract manganese from the subsoil and make it available to surface soil when the leaves fall and rot. Chlorosis is common in over-limed alkaline soil. Fortunately for us, manganese-deficient plants grow poor yields of food, and natural deficiency from foods probably does not occur unless they are over-refined.

There is adequate iron in most soils. When there is not, citrus fruits do not grow well and require additional soluble iron. A varied diet which includes meat should have enough iron to prevent anemia, but women, because of menstrual losses, can become anemic if they take strange, unphysiological diets low in this vital metal.

We have more concern with fluorine, cobalt, molybdenum, iodine and chromium. Soils deficient in fluorine, cobalt and iodine occur in wide areas. Large portions of the Mid-West are deficient in iodine, whereas areas near seacoasts are not, rains carrying enough inland for human needs. Because of excessive rainfall, many coastal areas are deficient in fluorine. Neither of these elements is essential for plants, and special supplements are necessary to avoid deficiencies in man. Likewise cobalt can be erratically distributed; much of the Eastern seaboard is deficient. Grasses grow well in cobalt-deficient soil and grazing animals become diseased. It is interesting that cattle and sheep will selectively choose grass grown where cobalt is seeded and avoid grass from a deficient area in the same pasture next to it, although differences in the two grasses are not apparent to the eye. Because the human requirement for cobalt is so small and because all of the cobalt used in vitamin B_{12} comes from animal sources, we can worry about human cobalt (B_{12}) deficiency only in strict vegetarians.

Molybdenum is required by legumes, for it is an essential element for the growth of nitrogen-fixing bacteria on their roots. These bacteria convert atmospheric nitrogen to soluble nitrates which are absorbed by the plants to synthesize proteins. Molybdenum-deficient soils occur where clover, alfalfa, beans and peas will not grow. The low concentrations of molybdenum, 1 ppm, on the earth's crust, accounts for the deficient areas.* Because human requirements are not known, we cannot form an opinion on natural deficiencies of molybdenum.

There is plenty of chromium in the earth's crust, but most of it is insoluble. Chromium deficiency in tissues is common in Americans and rare in Asiatics, Mid-Easterners and Swiss, from our analyses. There is considerable chromium in raw sugars, natural sugars, most grains but rye and corn, and vegetable fats. Chromium is largely removed in the refining of sugar and flour. There is much more chromium in wild animal tissues than in human. Whereas this subject is discussed at length in Chapter VI, at the present state of knowledge we have no evidence that natural chromium deficiency in man occurs. Human deficiency is probably almost entirely caused by food refining and fractionation.

Amounts in soils toxic to plants are found naturally in the case of copper, and toxicity to animals in the cases of molybdenum (in sheep) or selenium. It is doubtful that these natural toxicities are carried over in food to man. None of the essential trace elements is toxic except in very large amounts. In

---

*Molybdenum is a key element for life. It is essential for the bacteria which fix nitrogen from the air into a form, nitrates, which can be assimilated by plants to make proteins. It is essential for fungi, molds which live on dead matter, breaking it down to simple products used as nutrients. It is essential for blue-green algae, which fix nitrogen from air for solution in sea water as nitrate, or in soils, in lichens, ferns and mosses. Vanadium can substitute for molybdenum in some of these bacteria. If there were no ways to convert nitrogen in the air to nitrate, all life would slowly cease. But then, so would it cease if any other trace element essential for bacteria were missing.

fact, diets and supplements containing a tenth to all of the total body content of each of the essential trace elements can be tolerated daily by laboratory animals, except when the salt is irritating locally to the stomach.

We should mention briefly four other elements which occur in food and which may have biological functions: vanadium, nickel, arsenic and boron. Boron is essential for plants, and areas of deficiency in soils are found; pitted apples are a sign, cured by spreading borax under trees. As far as is known, mammals do not need boron, and excrete it readily in urine. Arsenic is ubiquitous and it has been found to take part in some reactions involving phosphate. Arsenic in its highly oxidized form is toxic only in quite large quantities but there is no good homeostatic mechanism other than urinary excretion and hair. Nickel has the requisite atomic structure for biological reactions; suggestions have been made that the color white in various living things may be conditioned by nickel, but we do not know. There is probably a good homeostatic mechanism in mammals for nickel. Vanadium depresses the synthesis of cholesterol and fat by rat liver and has been found to lower human serum cholesterol somewhat in some individuals but not others. It is concentrated highly in fats, especially vegetable oils. It is necessary for some nitrogen-fixing bacteria in the soil, like molybdenum, helps to prevent tooth decay in rats, and is required by some primitive forms of life, including the tubercle bacillus. Ascidians (sea squirts) have large amounts of vanadium, up to 4 percent, in their blood cells, which are green as a result. All indications are that it plays a biological role, but the definitive mammalian experiment has not been made.* A homeostatic mechanism for vanadium is probable. The amounts of these elements are shown in Table IV-1.

To sum up this chapter, we can say that there are nine or possibly ten essential trace elements, and perhaps two more. Unless foods are refined, a diet suitable for an omnivore, which man is, contains adequate amounts of seven: iron, manganese,

---

*Since this book went to press, vanadium has been found to be essential for the growth of rats.

chromium (if foods are unrefined), zinc, copper, cobalt, and molybdenum. Fluorine and iodine deficiencies occur as a result of geographical differences in soils. Information on selenium is incomplete.

CHAPTER V

Requirements, Intake and Excretion of Essential Trace Elements

Because of efficient homeostatic mechanisms for each element, which either prevent absorption of excesses, excrete them rapidly, or hold on to small amounts in the presence of deficiencies, there is usually a wide leeway between the lowest and highest intakes consistent with health. This large range of tolerable intakes was evolutionarily necessary when animals left the constant environment of the sea and walked on land. Because the distribution of essential trace elements is uneven over the crust of the earth, animals have adapted to reasonable excesses and partial deficiencies in many locations. When an essential element is present in toxic amounts, mammals do not normally survive.

Animals do not survive in areas where a trace element essential for life or for reproduction is missing or unavailable. However, if an element not essential for life but necessary for optimal health of a tissue or organ is low or absent, animals may survive in that area at reduced health, provided that absence of the element does not interfere with reproduction. We recognize two common examples of this kind of deficiency in human beings, in iodine, which prevents goitre, and in fluorine, which prevents decayed teeth and softened bones. In order to prevent goitre in areas deficient in iodine, we add sodium iodide to table salt, a practice which is 100 per cent effective. In order to prevent tooth decay and softened bones in the elderly, enlightened and knowledgeable people add sodium fluoride to their water supplies at levels below its concentration in sea water (1.3 ppm fluorine). This practice is 50-60 per cent effective on teeth, and presumably more effective on bone.

There are a number of examples of local deficiencies and excesses as reflected in diseases of domestic animals. Soils of pastures low in cobalt produce vitamin B_{12} deficiency in cattle and sheep feeding on them because of deficiency of cobalt in the grass and hay. Soils high in molybdenum transfer its toxicity to grasses which causes copper deficiency in sheep. Disease from selenium deficiency is well known.* Excess of selenium in pastures and especially in certain weeds concentrating this element produces brittleness of the hair and hooves and serious wasting disease, because selenium has the ability to replace sulfur in tissues (See Periodic Table). Moderate excess of arsenic apparently causes few changes, nor does partial deficiency. Fortunately for civilized man, the distribution of refrigerated and preserved food is such that diseases due to local excesses or deficiencies seldom appear, if at all; man is more tolerant of these large variations than are many animals.

In order that we have optimal health, we must consider the absolute minimal requirements, the amount we actually take in, the optimal amounts based on those needed by other mammals, and the levels that may be toxic. Table V-1 shows the amounts of the essential trace metals in representative American diets, in drinking water and other fluids, and in air. The airborne metals are inspired into the lungs, half being later driven up into the throat and swallowed, while the other half remains in the lungs. Chromium accumulates in the lungs in insoluble forms and is

*Selenium deficiency occurs in selenium deficient areas where levels in pastures are low. Still-births of live-stock and wasting of the young which survive can be prevented by adding small amounts of selenium to feed. Experimental selenium deficiency causes serious liver disease in rats and skin disease in chickens. Needs are very small.

TABLE V-1

Daily balances of essential trace metals

	Chromium μg	Manganese mg	Iron mg	Cobalt μg	Copper mg	Zinc mg	Molybdenum μg
Intake							
Food	100	3.7	13	290	3.5	13	280
Water	10	0.064	0.14	10	0.06-0.5	1	16
Air	0.03-0.3	0.002	0.027	0.1	0.02	0.012	0.1
Total	110	3.8	13	300	4.0	14	300
% Absorption	10	3-4	6.5	63-97	32-60	31-51	40-60
Output							
Urine	10	0.03	0.25	200	0.05	0.05	150
Feces	98	3.7	12	94	3.5	12	125
Sweat	1	0.039	0.5	4	0.04-0.4	0.78	20
Hair	0.6	0.002	0.013	0.04	0.003	0.03	0.01
Total	110	3.8	13	300	4.0	14	300
Retained in body	loss	0	0	0	0	0	0
Total body content, mg	1.4	12	4,200	1.5	72	2,300	9.3

Note: All of these essential trace metals increase in the body of man with growth and remain constant throughout life, except for chromium, which in Americans declines with losses induced by chromium-poor sugar (see Chapter VI). $1000\mu g = 1$ mg unavailable to the rest of the body, whereas the other six are absorbed into the body from the lungs. The per cent of metal swallowed in food and water which is absorbed into the body varies from less than one in the case of chromium to nearly all in the case of cobalt.

The metal not absorbed is excreted in the feces. That absorbed goes into the liver and sometimes into the pancreas; in the cases of chromium, magnesium, copper and zinc most is excreted into the intestine in bile. Urine is the principal route of excretion of cobalt and molybdenum. The well-absorbed elements cobalt and molybdenum are found in sweat in sizeable quantities, and under conditions of excess sweating large losses of elements in serum, such as iron and zinc may also occur. Hair represents a minor route of excretion of these metals.

We can get some idea of the relative amounts of a metal in food by comparing it with the total bodily content. In this way, we find that we eat every day 1.4% of the chromium, 14.6% of the manganese, 0.27% of the iron, 16.7% of the cobalt, 4.7% of the copper, 0.58% of the zinc, and 2.7% of the molybdenum in our whole bodies. When, however, we calculate the amount absorbed into the body compared to the total body content, which is what counts, we find that we absorb about 0.15% of the chromium, 0.5% of the manganese, 1.7% of the iron, 12.5% of the cobalt, 2.3% of the copper, 0.23% of the zinc and 1.4% of the molybdenum in our whole bodies. If these amounts were not excreted, they would rapidly accumulate. Therefore, low absorption of an essential trace metal makes it easier for the body to excrete what is absorbed and stay in balance. When absorption is relatively great, the kidneys put out excesses (molybdenum and cobalt).

In table V-2 are shown the same data for the essential non-metals. Most of the fluorine, iodine and boron are absorbed and excreted in the urine; more than half of the selenium is absorbed with a third in the urine and half in the sweat. Strontium is absorbed along with calcium, by the same mechanism, with some antagonism between them. When we calculate the amount absorbed compared to the total body

TABLE V-2

Daily balances of essential trace non-metals and one alkaline earth.

	Fluorine mg	Iodine μg	Boron* mg	Selenium μg	Strontium mg
Intake					
Food	1	200	1.1	150	1.8
Water	1.4	5	.23	?	0.1
Air	?	<50	?	?	?
Total	2.4	205	1.3	150	1.9
% Absorbed	80-90	100	99	60	17-38
Output					
Urine	1.6	175	1.0	50	0.34
Feces	0.15	20	0.27	20	1.3
Sweat	0.65	6	?	80	0.24
Hair	?	2	0.008	0.3	0.002
Total	2.38	205	1.3	150	1.9
Retained in body	0.02	0	0	0	0
Total body content, mg	2600	36	48	21	340
Organ of storage	Bone	Thyroid	Bone	all	Bone

*Essential for plants only

Note: Fluorine is retained in bone only when insufficient amounts are present.

contents, we find that 0.85% of the fluorine, 0.55% of the iodine, 2.6% of the boron, 0.43% of the selenium and 0.12% of the strontium, of the total enters the body. None of these is normally retained save fluorine in deficiency areas; when saturation of bony stores is reached, little or none extra is retained except under toxic and very high intakes.

These calculations do not tell us whether or not minimum requirements are being met by modern diets. All we know is that five specific elemental deficiency diseases have been described in man: chromium, iron, zinc, fluorine, iodine. Deficiencies of copper, manganese, cobalt, molybdenum occur only in experimental or domestic animals, in so far as is known. We can look, however, at the concentrations in human tissues according to age, and also compare human concentrations with those of wild animals, as a kind of natural control.

During such studies, it became evident that all essential trace metals except chromium were concentrated in the newborn and in infants and young children, levels declining in the second decade of life but remaining constant to the eighties; there were no examples of any organ lacking a metal. In the case of chromium, however, in American tissues deficiencies appeared in the teens and continued to appear throughout life, although infants and children had large amounts; thus Americans showed chromium deficiency of wide spread incidences. In some decades as many as 25 percent of Americans had no detectable chromium in tissues. In foreign tissues from the Middle and Far East there was virtually no chromium deficiency. The results of this deficiency are discussed in Chapter VI.

Iron deficiency is the result of chronic loss of blood or of some disease interfering with iron absorption. Because of the periodic loss of blood during the normal menses, women require more iron than men, especially when they are on diets poor in iron.

Zinc deficiency results in retarded growth and anemia. It has been found only in Iranian villagers and Egyptian oasis dwellers on very poor high carbohydrate diets in hot climates. It probably is fairly common in this country, especially in older people and pregnant women. Zinc is prevalent in all foods but refined flour, and sugar and fats.

Fluorine deficiency is wide-spread in this country, accounting for many if not most of the decayed teeth which plague the population and make mandatory so many dentists. Fluorine deficiency probably is also partly responsible for the large number of fractured hips in older people, which cause so much disability and even death. It is rare among people eating much sea food or drinking much tea; both contain considerable fluorine.

Iodine deficiency results in "colloid" goitre, those unsightly overgrowths of the thyroid gland seldom seen today in civilized countries, but formerly quite common in areas far from the sea, its principal source. Their disappearance is due to the wide use of iodized salt.

It is possible that deficiency of molybdenum also occurs but we know little about it. Dietary deficiency of cobalt is unlikely except in strict vegetarians; human requirements are 1 microgram of vitamin B_{12} a day, or 0.043 micrograms of cobalt, all of which must come from animal sources. Copper and manganese deficiencies are most unlikely in man.

What are the tolerable limits of the trace elements? Information of human tolerances are sketchy at best, for no one wants to cause toxicity which may be fatal (nor does anyone want to deprive a human being of an essential element to the point of causing disease). According to Bertrand's Law, every element is toxic in large doses, and an element is essential only when life cannot exist without it. I have broadened the definition of essential to include health and longevity, which to me is logical. We must turn to laboratory animals for comparative data. In table V-3 is shown what is known about minimal requirements, maximal tolerances and toxicities, reduced to human terms.

In respect to toxicity, we must remember that there are two kinds, metabolic or tissue toxicity from absorbed element, and local irritative effects on the stomach and intestine caused by

TABLE V-3

Minimal requirements, average oral intakes, tolerable limits and toxic quantities of the essential trace elements in terms of Standard Man and the rat, daily in milligrams.

	Minimum		Average		Tolerated		Toxic Amount	
	Rat	Man	Rat	Man	Rat	Man	Rat	Man
Chromium	0.98	0.20	2	0.21	49	10	>245	?
Manganese	1.7	1?	10	3.8	2550	>30	3400	?
Iron	2+	2+	340	13	?	2000	>4000	?
Cobalt	<0.001	<0.001	0.12	0.3	175	140	14,000	490
Copper	2	2	26	4.0	300	>15	>500?	?
Zinc	4	4	51	13	245	150	4250	?
Molybdenum	?	?	1.0	0.3	?	10	?	?
Fluorine	1	1	110	2.4	200	20	500	40
Iodine	0.1	0.1	2	0.1	1000	1000	>1000	>1000
Selenium	?	?	0.05	0.15	5	?	10	?
Strontium	?	?	?	1.9	?	4.9	?	?

Data shown for the rat are calculated for a 70 kg animal, in order that they be compared with man directly.

direct contact of the metal or its salts. It is the first that concerns us.

The table shows that in general the minimal requirement of each element in the rat and man is equal, that rats on commercial diets (and other laboratory and domestic animals) receive 2.5-8 times as much as does man, that very large excesses are tolerated without toxicity by both types of animal, and that human toxicities have generally not been established. Therefore, there is much leeway in larger doses of the essential trace elements, which attests to very efficient homeostatic mechanisms for repulsion of excesses and to their low toxicities.

There are six elements found in food and in tissues which are non-toxic orally to mice and rats in trace amounts, and for which biological functions have been suspected. Vanadium can suppress cholesterol and fat metabolism. Nickel may have something to do with color. Arsenic can affect phosphate energy exchange. Rubidium antagonizes the effects of vital potassium. Aluminum is usually added to laboratory animal feeds (why I don't know, but perhaps it was done once and became a habit). Bromine and iodine replace each other. They are shown in Table V-4. We know little about them now, and can only keep the subject open until more is known.

Is civilized man taking optimal quantities of the essential trace elements? We do not know, except in the cases of chromium and fluorine, but we can compare further the amounts fed to various animals with the amounts taken in by man, according to concentrations in dry diets (Table V-5). Compared to dog chow, human beings are receiving 1/36 percent of these elements; without copper the amounts are 1/17 percent. Because it is doubtful that man differs much in his needs from other omnivorous animals, we could build up a good, if very indirect, case that man is not getting enough manganese, cobalt, fluorine (which we know) or chromium (which we also know).

A better case, although still indirect, could be made for marginal intakes of essential trace elements resulting from the losses due to the refining of flour (Table V-6). The ash, which

TABLE V-4

Estimated daily balances of elements with possible biological functions.

	Vanadium mg	Nickel μg	Arsenic μg	Rubidium mg	Aluminum mg	Bromine mg
Intake						
Food	2.0	400	1000	1.5	45	7.5
Water	0.1	14	?	?	+?	?
Air	0.002	0.6	1.4	?	0.1	?
Total	2.1	415	1000	1.5+	45	7.5
% Absorbed	5	5	5?	90	0.1	99
Output						
Urine	0.015	11	50	1.1	0.1	7.0
Feces	2.0	380	900	0.3	43	0.07
Sweat	?	20	?	0.05	1.0	0.20
Hair	?	1	5	?	0.0006	0.002
Total	2.1	415	1000	1.5+	45	7.5
Retained in body	0	0	+	0	0.005	0
Total body content, mg	25	10	8	320	61	200
Organ of storage	Fat	Skin	Skin	All	Lung	All

TABLE V-5

Essential trace elements in human, rat, rabbit and dog diets, ppm

Element	Human	Rat	Rabbit	Dog	Human/Dog, %
Iron	30.6	197	252.6	200	15.3
Zinc	30.6	30.3	32.5	178.3	17.2
Manganese	4.62	54.4	44.4	59.95	7.7
Copper	6.15	15.1	13.0	17.05	36.0
Cobalt	0.1	0.37	0.38	0.48	2.1
Fluorine	0.59	65	—	50?	0.91
Iodine	0.12	1.17	0.59	2.25	5.3
Chromium	0.12	0.17	—	4.24	2.8

Note: Data on animal diets from Purina Laboratory Chows.

contains most of the elements, is reduced 75 percent. In white flour there is 13% of the chromium, 9% of the manganese, 19% of the iron, 30% of the cobalt, 10-30% of the copper, 17% of the zinc, 50% of the molybdenum, and 17% of the magnesium found in whole wheat. Iron only is restored to white flour, leading to potential deficiencies of all of the other elements. Just why iron is replaced and not zinc, which is just as important as iron, or chromium, or manganese, or magnesium, each of which plays a role in the metabolism of flour, is not known, but is probably one of those illogical and inconsistent decisions by committee, in which the opinions of the most cautious and unimaginative members prevail.

Not only are the essential trace elements removed from flour during its refining. There are large losses of the essential organic micronutrients also, the vitamins, especially of the B group, (Table V-7). In white flour there is 23% of the thiamin, 20% of the riboflavin, 19% of the niacin, 29% of the pyridoxine, 50% of the pantothenic acid, 33% of the folic acid and 14% of the vitamin E found in whole wheat. The first three are replaced. Remembering our logical postulate, that all natural raw foods contain the micronutrients necessary for their metabolism in sufficient quantities, one cannot help but wonder if refined flour contains enough of them to metabolize it

TABLE V-6

Essential trace elements in wheat and refined flour and in the fractions fed to domestic animals and poultry.

Element	Wheat	Flour	Germ	Millfeeds
Percent of whole wheat	100	72	2.5	25.5
Ash %	1.6	0.4	4.2	2.1-6.5
Chromium, ppm	1.75	0.23	1.27	2.18-2.22
Manganese, ppm	24-37	2.1-3.5	95-147	65-119
Iron, ppm	18-31	3.5-9.1	39-58	47-78
Cobalt, ppm	0.07-0.2	0.05-0.07	0.09-0.14	0.07-0.18
Copper, ppm	1.8-6.2	0.62-0.63	7.2-11.8	7.7-17.0
Zinc, ppm	21-63	3.9-10.5	101-144	54-130
Molybdenum, ppm	0.30-0.66	0.16-0.39	0.67	0.70-0.83
Selenium, ppm	0.04-0.71*	0.01-0.63	0.01-0.77	0.26-0.70
Strontium, ppm	0.48-0.86	< 0.5	0.5-3.6	1.03-3.79
Magnesium, %	0.09-0.12	0.013-0.021	0.2-0.25	0.23-0.38

*Selenium content of wheat depends upon area where grown. Representative samples are here given. Obviously, selenium is distributed throughout the whole kernel, unlike the other trace elements.

TABLE V-7

Vitamin losses in the refining of whole wheat.

Vitamin	Wheat μg/g	White Flour μg/g	Germ μg/g	Millfeeds μg/g	Vitamin tablet* Required mg/day	Vitamin tablet* mg
B_1 (Thiamine)	3.5	0.8†	22.0	17	1	10
B_2 (Riboflavin)	1.5	0.3†	5.5	4.2	1.2	10
B_3 (Niacin)	50.0	9.5†	80.0	150	10	100
B_6 (Pyridoxine)	1.7	0.5	12.0	7.2	2	5
Pantothenic acid	10.0	5.0	25.0	22	–	20
Folic acid	0.3	0.1	1.5	1.1	–	–
E (Tocopherol)	16	2.2	125.0	38	–	15 units
Gross energy, Kcal/g	4.4	4.3	5.1	4.7	–	–

*Theragran, high potency vitamins with minerals. As the average diet weighs 1.7 kg, to obtain μg/g or ppm divide value by 1.7 to compare with wheat.

†These three are replaced, along with iron. There are actually some 24 vitamins and bulk and trace elements partly removed from wheat during its refining to white flour; four are replaced.

Note: Millfeeds, which contain more of each vitamin than does whole wheat, and much more than white flour which humans eat, are fed to cattle and domestic animals, who must benefit greatly in many cases. Note the slight difference in calories; the higher level in germ comes from the oil.

properly. Mice and rats do badly on refined flour and well on whole wheat as major constituents of their diets, a fact which strongly suggests that the flour is deficient in almost everything but calories. It is no wonder that a "balanced" diet, which supplies these missing micronutrients from other foods, is mandatory in man and his animals, in order that they be healthy.

Therefore, we can strongly suspect that a modern diet obtaining a large proportion of its calories from white flour may be deficient in several micronutrients, especially pyridoxine, vitamin E, chromium,* manganese, zinc and magnesium. The consumption of grains in grams per day is 246 in the United Kingdom, 346 in Europe and 207 in the United States. Only in the United States is the standard flour refined to 72 percent of wheat; it is coarser in Europe and therefore contains more vitamins and trace elements. In terms of energy, this means that a third of a 2400 calorie diet may be deficient in one or more trace substances, and is almost certainly deficient in chromium.

Whereas about a third of our daily energy is supplied by grains, most of which is white flour, another sixth to a fifth comes from refined sugar.† In the United States, the refining of sugar has been developed to a more efficient level than in Europe, according to our observations and analyses. Whereas we do not know the whole story on sugar, we know that most of the ash and most of the chromium (92%) is removed during the refining process. Thus, another major source of calories, which in some people with a sweet tooth may amount to 25 percent, is depleted of a trace element necessary for its metabolism, and the result is a disease (see Chapter VI). The by-product of refining sugar is molasses, which is fed to farm animals and on which they thrive, as it contains all of the trace elements

*In 207 grams of whole wheat flour there are 362 μg chromium; in refined white flour 48 μg.

†The average consumption of sugar by Americans is 97 lbs per year or 0.265 lbs (120 g) per day. This amount supplies 424 calories, or 17.7% of a 2400 calorie diet. At 113 lbs per capita, a figure from a different source, calories from sugar amount to 20.1%. In 120 g refined white sugar are 3.6 μg chromium—or less. In 120 g raw sugar are 36 μg chromium.

TABLE V-8

Losses of trace elements due to the refining of sugar.

Element	Raw-brown	Refined white	Molasses
Chromium, ppm	0.24-0.35	0.02-0.35	1.21
Cobalt, ppm	0.40	< 0.05	1.26
Copper, ppm	1.34	0.57	2.21
Zinc, ppm	1.62	0.54	8.28
Ash%	3.2	0.11	8.0

removed from pure sugar. In Table V-8 is shown the magnitude of the losses.

Thus it is obvious that of the energy we daily expend and which is obtained from the burning of foods, up to half is strongly suspect as supplying insufficient trace elements for its efficient metabolism.

What can we do about this unfortunate situation, the result of habit and commercial interests? There are three things we can do:

1. For ourselves, we avoid the refined flours and all sugar-containing foods, using whole wheat, brown rice (which also loses elements in being polished white), dark brown sugar with a high ash content, and all the natural sugars: corn, maple, honey, fruit juices, etc. Much "raw" sugar imported into this country is considerably refined, because of excise taxes, and sugar with a high ash content is difficult to find. Brown sugar sticky from molasses is suitable; any sugar which flows freely is probably partly refined, regardless of color.

2. We can demand that those essential micronutrients, especially pyridoxine and chromium, which are removed from flour, be restored to it. To remove 13 or more (probably 18) micronutrients from wheat and to put back four is nothing short of illogical. We can demand that chromium be restored to white sugar in a form which can be absorbed by man.

3. We can demand that a coarser flour, including some low grades, be made available, which would contain at least 85-90 percent of the wheat or more (see tables V-6 and V-7). In England and Canada, flours of 70, 75, 80, 85, 90 and 100 percent of the whole wheat

were marketed. We can demand that true raw sugar, with at least 3 percent ash, is made available.

It is interesting to learn in what other foods the essential trace elements occur in worthwhile quantities, in order that we may balance our diet if we wish. In Table V-9 these data are shown, a do-it-yourself diet balancer. Iron was not included; values are available in any diet manual.

A careful look at this table reveals several interesting facts about trace elements in groups of foods. The highest values of chromium were found in nuts and cereals. Saturated animal fats had similar values, whereas vegetable oils appeared to have less affinity for chromium. Fats also were high in niobium, vanadium and copper. Nuts, being fatty foods, also contained chromium and these other elements. Raw sugar, molasses, fruit juices and whole wheat (but not rye) contained significant amounts.

Manganese occurred in high concentrations in nuts and leafy vegetables. Tea was a large source; an infusion had 6.9 ppm in the fluid. Nuts had 6.9-35.1 ppm. Whole grains were excellent sources, more than 10 ppm being found in wheat, buckwheat, rye and barley, as well as macaroni and grapenuts. From 4-8 ppm was contained in flour, oats, raisins, beet and turnip greens, spinach, rhubarb, cod liver oil and pressed linseed oil. Dried tea leaves had 276 ppm and ground coffee 20.7 ppm. In tea drinking countries a third or more of the daily manganese absorbed comes from tea. Sea food, meats, most fruits, legumes and fruity or root vegetables were poor sources, and we did not find any in dry skimmed milk—it was all in the butter. Therefore a "beef and bourbon" or "drinking man's diet" is low in manganese, which to be balanced would require much tea. All animal muscle, including human, was low in manganese. Without green vegetables, tea or coffee, a low carbohydrate, high protein, high fat reducing diet of 1500 calories could contain as little as 0.28 mg manganese, whereas 13-20 mg could be in a 2300 calorie vegetarian diet of whole grains, fruit, nuts, tea and fresh vegetables, well spiced.

TABLE V-9

Essential trace elements in various types of foods, average values ppm.

Type of food	Chromium	Manganese	Cobalt	Copper	Zinc	Selenium	Molybdenum
Sea foods	0.17	0.05	1.56	1.49	17.5	0.57	0.10
Meats	0.13	0.21	0.22	3.92	30.6	1.07	2.06
Dairy products	0.10	0.70	0.12	1.76	8.6	0.02	0.14
Vegetables							
Legumes	0.05	0.44	0.15	1.31	10.7	0.02	1.73
Roots	0.08	0.78	0.13	0.69	3.4	< 0.02	0.23
Leaves and							
fruits	0.03	3.47	0.14	0.42	1.7	< 0.02	0.06
Fruits	0.02	1.0	0.14	0.82	0.5	< 0.02	0.06
Grains and cereals	0.31	7.0	0.43	2.02	17.7	0.31	0.33
Oils and fats	0.15	1.83	0.37	4.63	8.4	–	0.00
Nuts	0.35	17.7	0.26	14.82	34.2	0.72	?
Condiments and							
spices	3.3	91.8	0.52	6.76	22.9	0.24	0.45
Beverages							
Alcoholic	–	–	0.03	0.38	0.9	–	0.08
Non-alcoholic	–	3.8	0.01	0.44	0.2	0.35	0.03

The largest amounts of cobalt were found in shrimp, scallops, smelt (7.2 ppm), blue gill and cod; other shell fish had small amounts. Canned fruits and vegetables had little (0.04 ppm). Concentrations of 1 ppm or more rarely occurred in over 200 foods: a baby food, bran, All-bran, sea salt (4.6 ppm), molasses, cocoa, and cat food. Five laboratory animal foods had 0.91 ppm, again attesting to the superior amounts of essential trace elements animals are fed, compared to man.

Large amounts of copper, 137 ppm, were found in oysters; clams had 3.3 and shrimp 3.4; other sea foods had less. Concentrations over 5 ppm occurred in liver, kidney, lamb chops, buckwheat, grapenuts, margarine (24.7 ppm), sunflower oil, vegetable and egg lecithins, all nuts, laboratory chow and sunflower seeds. Low values, less than 0.5 ppm, appeared in beef kidney, whole milk, a number of vegetables, coconuts, white bread, corn and brown rice. Distilled liquors showed some copper, probably as a contaminant from the still. Diets can vary several fold; low intakes are difficult to achieve and are monotonous, avoiding most meats and fish, eggs, fats, legumes, most vegetables, fruits, nuts, some grains, gelatin, tea, coffee, chocolate, soft piped water, soft drinks, beer and liquors. It is obvious that spontaneous copper deficiency in man from deficient diets is most unlikely. Therefore, we can more or less neglect copper.

Not much zinc is found in fruits and vegetables, except legumes. The richest source is muscle meats and fish. Oysters have contained 1074-1487 ppm zinc (one oyster pumps over 20 gallons of sea water through its gills in a day, and absorbs zinc, cadmium, copper from water). High values, more than 20 ppm, are found in egg yolk, wheat, gluten, rye, oats, buckwheat, corn flakes (76.5 ppm), lima beans, six nuts, tea leaves, cocoa, yeast. Low values, less than 2 ppm, occurred in butter, white of egg, white bread, polished rice, beans and peas (certain varieties only), many root and leafy vegetables, fats, fruits and sugars. Therefore we need whole grains, nuts, meats, and seafood in order to be certain that more than sufficient amounts of zinc are available. When a low zinc, high cadmium diet is taken for long, high blood pressure may develop (see Chapter VIII). Extra amounts of zinc would not be necessary if cadmium intake was low.

We know too little about selenium to estimate requirements. Deficiency of this element is partly relieved by vitamin E, and vitamin E deficiency is partly relieved by selenium.

Molybdenum is necessary for purine metabolism, and the foods in which it occurs illustrate our logical postulate that natural foods contain the micronutrients necessary for their metabolism. Purines and molybdenum are found in legumes and organ meats in large quantities. In man, molybdenum occurs largely in his liver and kidney, not in his muscle. Milk is also a source of molybdenum, where it is a constituent of an enzyme.

There is also little information on fluorine in foods, strange to say, for there are over 7000 references to it in the literature. Tea is an excellent source, as are seafoods; tooth decay seldom is seen in Caribbean Islanders, for example, as large amounts come from fish. Fishmeal, which contains the powdered bones, has much fluorine. Superphosphate fertilizers, which are mainly fossile fish teeth, are full of fluorine,* which may be transferred to food. The best way of getting enough fluorine, however, is in water.

We probably do not need to worry about strontium, except the radioactive kind. It is a normal constituent and may also prevent tooth decay and soft bones.

Spices can contain a relatively enormous amount of trace elements, but are poor sources as we eat so little of them. Copper is concentrated in thyme and black pepper (24 and 21 ppm), as is chromium (10 and 3.7 ppm) and manganese (83 and 47 ppm). Manganese is also concentrated in cloves (263 ppm) ginger (87 ppm), bay leaves (67 ppm), and tea leaves, as we have discussed. Cloves also have 1.5 ppm chromium and 8.7 ppm copper. One wonders if the piquant flavor of these spices is in

---

*Superphosphate fertilizers containing 14-33% calcium and 5-20% phosphorus have 2% iron and 2% fluorine. There is about 7-13 times as much fluorine as magnesium, and fluorine is the most abundant trace element.

TABLE V-10

Accumulation or need for essential trace elements during
pregnancy in maternal tissues, 40 week old foetus,
and mother's milk.

Element	Foetus	Mother	Milk, loss/day
Chromium, μg	660	800	?
Manganese, mg	1.0	1.5	.006
Iron, mg	—	—	1
Copper, mg	14	23	0.38
Zinc, mg	43	130	5
Magnesium, mg	730	1300	30

some way related to the tastes of organic complexes of these
necessary trace elements.

Pregnancy represents a special situation in respect to trace
elements. All of the essential ones are concentrated in the
growing embryo, and at birth relatively high concentrations are
found. These decline during the first year or decade of life,
remaining constant thereafter. The growing foetus demands
trace elements from the mother's tissues. If not in adequate
supplies, the demands of the foetus deplete the mother. In
Table V-10 are shown the needful accumulation of elements in
the 40 week foetus at term, and in the mother's tissues, and the
daily loss in milk of the lactating mother. One can see that
relatively much zinc and chromium are required. During pregnancy, the mother's diet must contain more than enough to
cover her own and her baby's needs, of 1.5 mg chromium and
200 mg zinc.

All of the data in this chapter, and all of the words, have
merely demonstrated which essential trace elements we do not
need to think about in our diets—cobalt, copper, molybdenum,
selenium, strontium—which ones we should be careful about
when we change the proportions of different foods on special
unphysiological diets—manganese, zinc—which ones we need
under special circumstances in certain geographical areas—
iodine, fluorine—and which one we need anyway, for we are

probably deficient—chromium. This chapter has also told why. I hope that it has clarified, and not confused the problem, in spite of so many figures.

CHAPTER VI

Chromium Deficiency and Atherosclerosis

Atherosclerosis, or hardening of the arteries, is a very common disease of civilized man, varying in severity and incidence around the world, accounting directly or indirectly for much of the disability and about half the death rate of older people in the United States. It is rarely seen in primitive man but ancient Egyptians had it. For many years this disease, because of its frequency, was believed to be an invariable accompaniment of aging—from which belief rose the saying "A man is as old as his arteries." Surveys of incidence and severity in different countries during the past quarter century, however, have shown the fallacy of this reasoning, and atherosclerosis is now recognized as a disease occurring in epidemic proportions in this, and a few other countries.

It is a disease of the arteries, characterized by fatty deposits under the innermost layer. These deposits produce plaques or irregularities in arteries, interfering with the smooth flow of blood. At first they are streaky, then raised and lumpy. The surface may break down over the lump, leading to a clot on it. When the artery is small, sometimes the whole artery is filled with clot. Healing may be associated with a ragged appearance and deposits of calcium. The largest artery, the aorta, which is normally elastic, may become rigid. Sometimes these bony plaques rupture.

All tissues need plenty of oxygen-carrying blood, and when an artery is narrowed, the part it supplies gets less blood than it needs. When one is clotted, the part gets none and can die. The damage caused by atherosclerosis thus depends on which artery is narrowed. In the heart, coronary occlusion—"heart attack"—results. In the brain, there may be a variety of manifestations: a stroke involving one side of the body, deterioration of the thinking process ending up in a vegetative existence, discrete disorders of the body from damage to discrete and specialized areas, such as dizziness, ringing in the ears, loss of speech, loss of vision, numbness in certain places—in fact, as many manifestations as there are functions of the brain. If the artery going to the kidney is narrowed, high blood pressure develops. If a leg artery is narrowed, the muscles ache with exercise—and such narrowing can lead to gangrene and loss of the leg. When the great artery—the aorta—is stiffened by disease, the systolic blood pressure goes up—much as a cold steam radiator will bang if there is no elastic cushion of air or steam in it. Arteries, however, have a marvelous ability to grow around an obstruction, opening up a "collateral" circulation. Therefore, it is only when the atherosclerotic process involves a major artery, and advances more rapidly than collaterals can grow, that the organ or part of the organ suffers from lack of blood. We do not know why the arteries of the heart muscle are more likely to be affected in some people and in some areas of the world, and the arteries of the brain in others.

Aside from these changes in the walls of arteries, there are other manifestations found by examining the blood. Fats and fatty substances, of which cholesterol is the most popular to measure, are increased above "normal" levels. There are signs of mild diabetes, in that a test of blood sugar is normal in the fasting state but elevated abnormally after a breakfast of sugar.

Researchers disagree on what is a "normal" blood cholesterol. During my thirty-five years of clinical experience, I have seen the "normal range" of blood cholesterol rise every decade, as the average increased in the general American population. Obviously this average was not normal, some influence acting to raise it. A comparison of "normal" levels in other countries of the world have generally shown low (or probably normal) levels in the blood of Indian laborers, African tribesmen, low income Bantu, American Indians in the Southwest, Thai, and other such people in many areas of the world. The fact that healthy human beings have blood cholesterol levels which are low compared to people exposed to Western Civilization indicates that the latter have a disorder in respect to this substance.

Cholesterol is a vital substance in the body of animals. Some organs, such as brain and pancreas, contain as much as five percent cholesterol, and the liver more than one percent. Without it the skin would dry up, the brain would not function and there would be no vital hormones of sex and adrenal. Just why and how it collects in the blood to deposit in the arteries is not known, but something is wrong with its metabolism.

Children in all societies have low levels of blood cholesterol, around 100 milligrams per 100 grams of blood, more or less. During adolescence the average level rises somewhat. At this age a dichotomy appears, which although not absolute, is evident in serial studies made in many areas of the world. Primitive and low income people living on ancient foods tend to keep adolescent levels of cholesterol (120-150) all their lives, whereas Westerners increase their levels with every decade of life until the sixth or seventh. Again this suggests that Western Man is diseased. This rise with age occurs in people of other races who have adopted Western ways, particularly in diets. There is little doubt that a major, if not *the* major factor in high blood cholesterol, high blood fats, and other abnormalities in the blood of people with atherosclerosis, is the type of diet common to the area.

For several years a theory prevailed that a high dietary intake of cholesterol was responsible for high blood levels and for atherosclerosis. This idea, however, proved false, except where intakes were extremely large. When cholesterol is eaten in egg yolks, animal fats and organ meats, and absorbed from the intestine, the liver makes less cholesterol and excretes more, so that there is little net effect. In no cholesterol (all vegetarian) diets the body still makes it. Only during very high intakes such as would be provided by a dozen eggs a day does the blood level rise. There appears to be a kind of cholesterol-stat, a regulator, in all people, which is set high in Western Man.

Then the theory that a high fat diet was responsible for high cholesterol—and severe arteriosclerosis—became extremely popular. We call this the Mrs. Jack Spratt theory. This idea was then modified to put the blame on "saturated" fats, that is, fats all of whose carbon atoms contained two hydrogen atoms, as opposed to "unsaturated" fats, some (1-4) of whose carbon atoms had only one hydrogen atom. (A fatty acid is a chain of carbon atoms of varying lengths. Natural ones have an even number of carbon atoms, from four to eighteen. Three fatty acids linked to glycerol make a fat. In general, fats solid at room temperatures are saturated, liquids are unsaturated. Saturated fats come mainly from animal sources, unsaturated from fish and vegetable sources. Coconut oil, however, is saturated. Fish have to have unsaturated fats or they would be solid at low water temperatures, while more solid fats liquid at body temperatures occur in animals. Man has learned to saturate many common vegetable oils, so that the housewife can spoon and melt them.) Giving very high doses of unsaturated fats to human beings will lower the blood cholesterol moderately, whereas the same doses of saturated fats will raise it. These experiments can hardly be considered physiological, as the doses used are enormous.

The Mrs. Spratt theory does not fit all the facts. For example, a 60 year old Masai living mainly on meat, animal fat, milk and butter (mostly saturated) can have a cholesterol of 140 milligrams per 100 grams of blood. A Trappist monk of the same age on a vegetarian diet plus eggs and milk can have a cholesterol of 190 milligrams. A Thai eating unpolished rice and fish can have 140. A Japanese living on polished rice, vegetables and fish and very little fat will show 165. An American on a usual Western diet, obtaining more than half of his calories from sugar and saturated fat will show 270; put him on a moderately low fat diet, mostly unsaturated fats, and his level may decline to 232 milligrams. The Arctic explorer, Vilhjalmur Stefannson, and a partner lived for a year on an all-meat diet with about 80 percent of his calories coming from fat; his blood cholesterol rose moderately at first and then dropped to a level below the original, i.e. 192. His partner, who later was found to be one of those persons always carrying a high cholesterol, showed a level of 600 or more; this man was specially susceptible, unlike most of the population. Unfortunately, his data are usually quoted, not Stefannson's.

Comparisons of the incidences of mortality from athero-sclerosis in various countries of the world and the per capita consumption of fat show some correlation when *one* manifestation of the disease is considered—coronary heart disease. This relationship becomes less if the major manifestations, heart and brain disease (stroke) are included. The underlying disease is atherosclerosis, a single entity striking at various points in the body or existing with no occlusion of an artery. In order to be exact, one would have to determine the incidence of all atherosclerosis and its manifestations compared with consumption of fat by country; this is difficult to do. One extensive study showed that atherosclerosis of the aorta among Japanese, obtaining nine percent of their calories from fat, largely unsaturated, was as severe as that among New Orleans natives, obtaining 40-45 percent of their calories from fat, mostly saturated. Both groups had significantly higher degrees of disease than found in Guatemalan natives living largely on vegetable and fruit diets. Strokes and coronary heart disease are common in New Orleans, strokes are much more frequent in Japan, but not heart disease, and both manifestations of this disorder are unusual in Guatemalans. These findings cast serious doubt on the validity of the Mrs. Jack Spratt hypothesis, which has changed the eating habits of a nation.

A comparison of mortality from coronary disease in various countries with the per capita consumption of carbohydrates (sugars and starches) and especially with that of sugar, is as valid as is that with fat, perhaps more so. Again, all manifestations of atherosclerosis were not included. The finding that practically everybody with clinical atherosclerosis of moderate severity has a mild form of diabetes, and the long known fact that people with moderate and severe diabetes have especially severe atherosclerosis, from which most of them die, links the two disorders of fat metabolism and sugar metabolism together and demands a search for a single causal factor basic to both.

In 1959, Klaus Schwarz and Walter Mertz, who had been studying a dietary deficiency in rats manifest by a reduced tolerance to glucose (mild diabetes), after an exhausting search found that the deficient factor was chromium. Some of the commercial diets had little of this trace element and caused the disorder; others had much and did not. The addition of chromium compounds to the diet cured and prevented the disorder. Their pioneering discovery in retrospect was to have one of the most important and far reaching effects on the nutritional aspects of human disease in the mid-20th century. Later Mertz showed conclusively that chromium was necessary for the utilization of insulin in glucose metabolism; a deficiency of available chromium or a deficiency of insulin had similar effects; an excess of insulin without chromium, or an excess of chromium without insulin gave like results, that is none. He also showed that chromium was necessary for insulin to cause the build-up of fat from simple substances.

Five years prior to this, George L. Curran made a highly significant discovery. Studying the effects of the transitional metals on cholesterol and fatty acid synthesis, he found that vanadium was strongly inhibitory, whereas manganese and chromium stimulated the formation of both substances by the liver. The rats he used were on a commercial diet then (but not now) deficient in chromium. If the rats had not been deficient, probably no effect of added chromium would have been seen.

In 1960, we began an exhaustive study on the effects of trace elements given to rats and mice from weaning until natural death, in small doses. Chromium, cadmium and lead were the first three given in this way; the controls received no chromium and the diet contained little. We were not aware at that time of how essential chromium was to health. Four years later the last of 104 rats receiving chromium died, somewhat of a record for longevity. They grew faster, survived longer, and at death, surprisingly enough, had no atherosclerotic plaques in their aortas, unlike their controls of which about 20 percent of aortas had plaques. These and other animals, about 700 in all, had elevated blood cholesterol and sugar levels when they were in an environment and on a diet low in chromium and low levels when chromium was given in addition. Special diets lower in chromium produced greater discrepancies when chromium was or was not added. Taking all of these findings together, we could believe that we had reproduced in the rat the human disorder in fat and sugar metabolism which leads to atherosclerosis. The degree of severity of the disorder, however, was mild to moderate; the rat is one of the most difficult animals to make atherosclerotic, requiring extreme measures; we had done it without extreme measures, using ordinary foods, merely by inducing chromium deficiency, and prevented it with chromium. Other studies further corroborated this idea. Rats were given a diet of half white sugar, a third torula yeast (which is low in chromium) and the rest lard. When chromium was given, cholesterol was low and mild diabetes absent; without it, cholesterol and blood sugar were elevated. When dark brown sugar replaced the white, both substances were low. Brown sugar contains six times as much chromium as does white.

Another interesting change occurring in chromium deficient old rats is the development of opacities of the cornea of the eye. There is a concentration of chromium in the cornea, as in the skin; opacities and vascularization were found quite frequently by Mertz and by us. No cataracts appeared; diabetic and atherosclerotic subjects are prone to develop cataracts (of the lens).

How do these findings fit into the human disease? It would be necessary to show that a sizeable percentage of people on Western diets are deficient in chromium, their bodies containing less than the bodies of other people. This difference was conclusively demonstrated by the analyses of Isabel H. Tipton on human tissues from different areas of the world, according to age. When we compared the concentrations of chromium in the tissues of Americans and foreigners at all ages, we found high levels in stillborn, newborn infants and children up to ten years of age, which declined precipitously in the next two decades. Chromium, present in all young bodies, was not detected in 15-23 percent of American tissues from people over 50, but was found in almost every foreign one (98.5%). Estimates based on organ weights indicated that Africans had twice, Near Easterners 4.4 times and Orientals five times as much chromium in their bodies as did Americans. (See Figures VI-1-3). While we cannot prove that those persons deficient in tissue chromium had severe atherosclerosis, we found that chromium in the aorta was not detected (too low to be found) in almost every person dying of coronary artery disease, one manifestation of atherosclerosis, and was present in almost every aorta of persons dying accidentally.

Additional evidence that chromium is low in American adults can be inferred from analyses of wild and domestic mammals, all of which were presumably healthy and none of which required a balanced diet (table VI-1). Average chromium was 21.4 micrograms per 100 grams of wet tissue in liver, kidney, heart, spleen and muscle of beef, lamb, pig, rats, squirrels, fox, beaver, woodchuck and deer. The same organs of Americans had 2.4 micrograms or about a tenth as much. It would be unreasonable to believe that chromium was essential for all mammals but man, so basic is its action.

Unfortunately, chromium in the forms available for feeding in pills to human beings is poorly absorbed by the intestine. Only about half of one percent enters the body. Chromium in the forms occurring naturally in foods is obviously better absorbed, for a small amount is found normally in the urine. No one at this writing knows its state in plant or animal tissues. Mertz, however, has grown yeast on increasing concentrations of chromium, and has extracted from it a well-absorbed organic complex of chromium, some 100 times more active than its

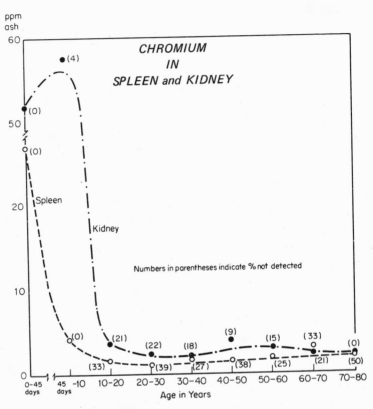

Figure VI-1: *Mean concentrations of chromium, when present, in 135 spleens and 176 kidneys of American subjects from 10 cities according to age. Means of all adults were 0.5 and 1.0 ppm, respectively. Renal levels at birth were maintained through the first decade; splenic levels were not. Numbers in parentheses show the percentage of samples in each age group which were deficient in chromium (< 0.1 µg/g ash) and not included in mean values. Mean splenic values of adult foreigners by area varied from 2.2-5.5 ppm with 2 of 144 deficient; mean renal values from 3.3-24 ppm with none of 154 deficient.*
From Schroeder, H.A., Circulation 35: 570, 1967, with permission of the publishers.

simple salts. As soon as Mertz' high potency chromium is available, much more will be learned about its effects on the human syndrome of atherosclerosis and mild diabetes.

Even the low potency chromic chloride and chromic acetate have some effects, although they are inconstant. On glucose

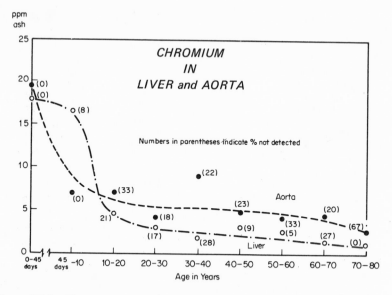

Figure VI-2: *Mean concentrations of chromium in 186 livers and 103 aortas of American subjects from 10 cities according to age. Mean values for all adults were 0.7 and 2.0 ppm respectively. The numbers in parentheses show the percentages in each age group which were deficient in chromium (< 0.1 μg/g ash), and which were not included in the mean values. Hepatic levels at birth were maintained for a decade before declining. Foreign tissues had higher values, livers showing means of 1.3-5.8 ppm, with 1 of 154 deficient, and aortas showing 6.6-30 ppm ash with 4 of 100 being deficient.*
From Schroeder, H.A., Circulation 35: 570, 1967, with permission of the publishers.

(sugar) metabolism, Glinsmann and Mertz found improvement of the diabetic state in three of six patients carefully studied in a metabolic ward; we found improvement in four of 12 outpatients; Levine and his colleagues found improvement in four of ten atherosclerotic patients; Hopkins found improvement in four of eight middle aged volunteers; these results are encouraging but not entirely conclusive. In children suffering from protein deficiency, the ability to digest glucose was restored to normal by chromium, provided the children came from an area where there was little chromium in the drinking water. Therefore, some effects of poorly absorbed chromium has been demonstrated on some, but not all human beings; in

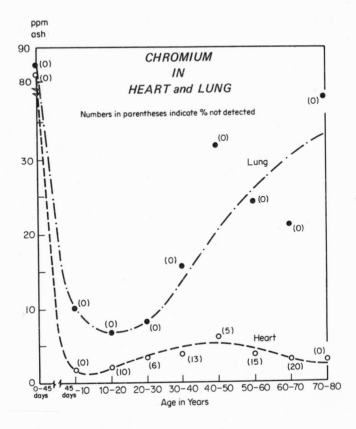

Figure VI-3: *Mean concentrations of chromium, when present, in 145 lungs and 140 hearts of American subjects according to age, Mean values of all adults were 20 and 1.5 ppm respectively. The numbers in parentheses are the numbers of samples in each group deficient (< 0.1 µg/g ash) in chromium. Note that all lungs contained this metal. The usual high concentration is seen in the newborn, characteristic of all essential trace metals studied. The rise in pulmonary chromium with age ($r = 0.39$, $P < 0.001$) is probably the result of deposition of airborne chromium as an industrial pollutant; note that this depot is apparently fixed and high concentrations are not reflected in other tissues. Fewer hearts were deficient than were other tissues. Foreign hearts had higher concentrations as follows: all of 43 Africans, 2.1 ppm; all of 20 Near Easterners, 5.6 ppm; 61 of 62 Orientals, 6.3 ppm ($P < 0.001$); all of 8 Swiss, 4.6 ppm. There were no significant differences in foreign lungs, 44 Africans having 17 ppm; 33 Near Easterners having 27 ppm; 69 Orientals having 32 ppm; and 7 Swiss having 62 ppm.*
From Schroeder, H.A., American Journal of Clinical Nutrition 21: 230-244, 1968, by permission of the publishers.

TABLE VI-1

Chromium in wild and domestic mammals and in subjects from the United States, mean values, wet weight.

	Animals µg/g	Human subjects µg/g
Liver	0.16	0.02
Kidney	0.18	0.03
Heart	0.14	0.02
Lung	0.24	0.20
Spleen	0.48	0.02
Muscle	0.11	0.03
Stomach	0.04	0.03
Placenta	0.07	0.42
Mean	0.19	0.05

Note: Animals included beef, lamb, pig, wild rats, squirrels, fox, beaver, woodchuck, deer; analyses by chemical methods. Placenta excluded from mean.

children the effects appear quickly, but in adults they take several months.

On cholesterol levels in the blood, we found lowering of 14.2 percent in five institutionalized patients after five months of chromium; in one outpatient it fell 26 percent in seven weeks. In another five, treated at the same time, changes were only 1.7-7.3 percent. When considered together, it is evident that the administration of chromium has a favorable, although mild to moderate effect of reversing those measurable blood changes found in atherosclerosis. These results suggest that the rat experiment might be duplicated in man if the right form of chromium were found.

The causes of the losses of chromium from body stores in adolescence and later are probably explained somewhat as follows: when pure sugar (glucose) is fed in a large amount to a fasting person, three changes take place: a) Blood sugar is elevated; b) Insulin in blood increases; c) Chromium in blood increases, being mobilized from tissue stores. In a diabetic, variable amounts of sugar are excreted in the urine; in normal people, little or none may appear. The chromium in the blood, elevated in response to the sugar and perhaps accompanying the insulin, reaches the kidneys where some is excreted, perhaps 20 percent of that in blood. If the sugar contains chromium, as all unrefined sugars do, presumably enough is absorbed into the body to account for urinary losses and there is no net loss. If the sugar contains very little chromium, there is a net loss from the body. One can calculate from actual data that 150 grams of glucose or white sugar containing 3-4 micrograms of chromium would cause a net loss of 12-25 micrograms of chromium in the urine in a day, or about 8.75 milligrams a year, more than the total body content; necessary chromium would have to be supplied from other foods. If the sugar had contained, on the other hand, 36 micrograms, as does raw sugar, there would probably be no net loss and the body would be in chromium balance (table VI-2).

We do not know what effect carbohydrates other than sugars might have on chromium balance, but it is reasonable to believe that low chromium starches may show the same effects. Starches are digested—or split—into sugars in the intestine and so absorbed. The major source of carbohydrate calories in the American diet other than sugar is refined white flour. Whole wheat contains 175 micrograms of chromium per 100 grams; refined white flour contains 23 micrograms. White bread has 14, whole wheat bread 49 micrograms (Table VI-3). Therefore, it is likely that white flour may cause depletion of body chromium, just as does white sugar. Table VI-4 shows the variations in the balance of chromium in ordinary diets and how easily net losses from the body can occur.

Fats also have varying amounts of chromium, depending on their types. The vegetable oils contained 4-17 micrograms per 100 grams, butter 21, hydrogenated vegetable shortening 16, margarine 14, eggs 17, lard 7. There was a rough relationship between the degree of unsaturation of the fat and the amount of chromium. Vegetable oils provided less chromium than animal fats.

Therefore, the typical American diet, with about 60 percent

TABLE VI-2

Chromium in sugars

	μg/g	μg/100 Cal.
Sugar cane, Virgin Islands	0.07	—
Sugar cane, Puerto Rico	0.12	—
Bark	0.15	—
Raw sugar, Philippines	0.24	6.0
Raw sugar, Colombia	0.35	8.8
White sugar, U.S.	0.08	2.0
White sugar, France	0.13	3.3
White sugar, Superfine, U.S.	0.02	0.5
Brown sugar, dark	0.12	3.0
Brown sugar, light	0.06	1.5
Brown sugar, Irish	0.07	1.8
Fructose	0.18	4.5
Glucose	0.03	0.8
Lactose	0.17	4.3
Cerelose	0.17	4.3
Molasses, household	0.11	4.3
Molasses, blackstrap	0.22	8.6
Molasses, refinery	0.27	10.5
Affination syrup	0.75	29.2
Molasses, final	1.21	47.0
Honey, purified	0.29	10.0
Maple syrup	0.18	6.0
Corn syrup	0.15	5.0
Orange juice	0.13	34.0
Grape juice	0.47	94.0

Data shown are means, obtained by atomic absorption spectrophotometry. White sugar varied from 0.04-0.10 μg/g in 10 samples.

of its calories from refined sugar, refined flour, and fat, most of which is saturated, was apparently designed not only to provide as little chromium as feasible, but to cause depletion of body stores of chromium by not replacing urinary losses. Again, the "article of faith based on reason" stated previously, that whole foods contain the micronutrients necessary for their metabolism, has been shown to hold true, and the refining of these foods based on custom, habit, preference and industrial

Note well

TABLE VI-3

Essential trace elements in various fractions of wheat used for human and animal consumption, ppm dry basis.

	Cr*	Mn†	Fe†	Co*	Cu†	Zn*	Mo†
Wheat, whole	1.75	49.0	37.3	0.75	4.0	31.5	0.33
Farina	0.58	4.6	5.4	—	1.6	6.6	0.17
Patent flour	0.23	6.0	7.4	0.36	1.5	8.9	0.19
First class flour	0.60	9.0	20.4	0.26	2.7	15.9	0.32
Low grade flour	2.19	35.8	43.6	0.75	5.5	33.6	0.39
Germ	1.27	137.4	66.6	0.50	7.4	133.6	0.67
Red dog	0.15	121.4	131.4	—	14.2	3.6	0.70
Shorts	2.22	164.7	145.7	—	13.3	106.0	0.79
Bran	2.18	136.5	141.3	1.24	15.2	100.2	0.83
Bread, white	0.14	5.9	27.3	0.11	2.3	1.2	0.32
Bread, whole wheat	0.49	41.3	—	—	5.1	5.3	—

Note: The fractions shown are those of commercial flour mills. The less refined fractions are usually used for animal feeds: low grade, red dog, shorts and bran.

*Analyses by atomic absorption spectrophotometry on samples from Peavy Co., Buffalo, New York.

†Analyses by colorimetric methods reported by Czerniejewski, et al.

TABLE VI-4

Approximate daily balance of chromium in man, microgram/day

	Intake	Output	
Food	100(5-500)	Urine	20(3-160)
Fluids	10(0-170)	Feces	30(10-370)
Air*	(0.03-0.3)	Sweat	1
		Hair & nails	0.6

*Not available as remains in lung.

Net urinary loss occurs on lower intakes if chromium-poor sugars, starches and fats are eaten. About half of the ingested chromium is absorbed, of which about 15 percent is excreted back into the intestine in bile.

practices has been shown to provide foods lacking in a sufficient amount of a micronutrient necessary for their metabolism. The result is a prevalent disease, in this case, atherosclerosis.

What can we as individuals do about this situation for ourselves? It is too late for most of us to prevent the depletion of our body stores of chromium, but we can prevent further depletion by wise dietary habits. And we may hope to build up our stores, in order that sugars, starches and fats be efficiently digested and burned. There are three ways to do this, although none has been proven experimentally. 1) We can avoid refined white sugar, and all foods containing it. These include jellies and jams, candies and sweet cakes, cola drinks containing glucose, pies and other sweet manufactured products. If we need to obtain 20-25% of our calories from sugar, we can use sugars high in ash, such as dark brown sugar (many imported raw sugars are actually quite refined), corn syrup, maple syrup, honey, fruit juices (Not those with synthetic flavorings and glucose), molasses (which contains all of the chromium removed from the raw sugar), and other untreated sugars. A no-sugar diet, using artificial sweeteners, would deprive us of needed calorie energy and should only be used for weight reduction. 2) We can avoid refined white flour, although it is difficult, substituting whole wheat and other whole grains whenever possible, unpolished rice, whole rye, oats, corn, etc. 3) We can avoid saturated fats as much as possible, substituting unsaturated vegetable oils. As to fish and meats, shell-fish and chicken contain much chromium; meats have several times as much as vegetables, grains and fruits.

Such diets are not unreasonable. 1) In this way, further depletion may be avoided and repletion may occur. The net loss of only one microgram of chromium per day from the body will result in a total loss of nearly 11 milligrams in thirty years, or about the total bodily content of Japanese adults; contrariwise the net gain of one microgram a day for thirty years would ensure adequate stores in old age. 2) Alternatively, we can take one or two milligrams of trivalent chromium in pill form a day, realizing that only about 50 micrograms would be absorbed. The proof of repletion of body stores should be evident from the blood cholesterol level and the test for sugar tolerance. 3) Or we can wait for high potency chromium complexes to become available. 4) Or we can take continuously one of a number of drugs which lower blood cholesterol. All of these drugs have side-effects, mild or serious. Any chemical substance designed by man and not Nature has late toxicity for a certain percentage of people when taken continuously. For example, one such agent prevented the final conversion of cholesterol's precursor, desmosterol, into cholesterol. The result of accumulation of desmosterol was the formation of cataracts of the eye. Certain metal binding agents will lower cholesterol; nicotinic acid, a B-vitamin, in large doses is one, as are two of the most common "chelating agents," which transport and remove trace metals from the body. A derivative of thyroid hormone is another. Whether or not these reverse the atherosclerotic-diabetic syndrome is not known. It is tempting to speculate that some of these drugs mobilize chromium from tissue stores where it is strongly bound.

According to this theory, the most efficient and the fastest way to develop atherosclerosis is to drink much coffee all day with three or four spoonfuls of sugar and cream; use marmalades and jams on breakfast toast, with sugar thickly spread with cream on processed breakfast foods; eat a couple of ham sandwiches for lunch on white bread, followed by a slab of pie; and for dinner take large portions of pork with rich gravies, deep fried potatoes, and a lemon meringue pie for dessert, followed by a sticky sweet liqueur or two. Such a diet will almost surely result in an elevated serum cholesterol, a depletion of chromium, and atherosclerosis.

What are the discrepancies in this theory that chromium deficiency is the basic cause of atherosclerosis? There are none to my knowledge, although all the holes have not been filled. Countries where incidence and severity of atherosclerosis is less than in the United States are those where unrefined flour is usually consumed, and those where sugar is less refined than by the marvelously efficient American methods. Countries where atherosclerosis is rare are those where major calories come from unrefined grains without much sugar. By knowing the dietary habits of a nation in respect to refined carbohydrate foods, and knowing the chromium content of the foods, and in some cases knowing the chromium content of adults' tissues, one can predict where atherosclerosis will be prevalent, where it will be severe and where it will be virtually absent. One can also predict what effect dietary changes will make in this disease, such as those which took place after World War II, when refined flours and sugars became available after wartime scarcities.

If what I have said is valid, and chromium deficiency turns out to be the major culprit in atherosclerosis, as all studies indicate that it is, we may eventually expect vast changes in the public health as regards disability and mortality from athero-sclerosis. These changes will take place slowly, for they will depend on whether or not repletion of chromium causes the disease a) to cease progressing, b) to slowly heal when mild to moderate lesions in the arteries are present, and c) to slowly heal advanced lesions. Because the major serious and fatal effects of atherosclerosis are usually caused by clotting of the blood in the artery at the site of a lesion, there would be no clotting if there were no lesions, but if there were lesions which did not heal but did not progress, the artery would still be susceptible to clotting. There is no evidence that chromium deficiency causes, nor chromium prevents, clotting in itself. Therefore the most effective way of avoiding the athero-sclerotic-diabetic syndrome is by prevention. To this end I look to the day, although I may never see it, when this essential micronutrient so necessary to the integrity of arteries is either not removed from our major sources of calories by refining, or if the public insists on white flour and white sugar, is restored in sufficient quantities to those foods to ensure their adequate metabolism. Modern man makes many mistakes through lack of knowledge, but there is no excuse for his continuing his mistakes in the face of knowledge.

There is another micronutrient possibly concerned in this disease. Vitamin B_6 deficient monkeys get lesions in the walls of their arteries similar to arteriosclerosis. When they are fed cholesterol, these lesions fill up with fat. Therefore, older persons should take enough B_6, probably 5 mg a day, to prevent this possibility occurring, as it may be the first step in the disease.

CHAPTER VII

The Non-Essential
Trace Elements

All the elements are found on the earth's crust, in soil and in sea water from which life began. They occur in concentrations from the lowest amounts detectable by sophisticated instruments and methods to quantities making up sizeable proportions of the media in which they are found. Although the line between what is a bulk element and what a trace element is arbitrary, and according to our definition, depends upon biological concentration, many elements classified as "trace" are unsuitable in structure and in chemical characteristics to take part in living processes. Life goes on with or without them, and they can be considered as inert contaminants of the natural environment since life began.

During life's evolution from single cells to man, living matter has utilized those elements which were a) ubiquitous in fair-sized quantities, b) reactive enough to take part as catalysts, in order that protein be built and carbohydrates and fats built and burned at low temperatures, and c) non-toxic in the amounts found in the environment. Therefore, we can expect that an element found in very small amounts in sea water, or one which precipitates to the ocean floor when it is carried from land to sea, will not be essential for marine (or animal) life, regardless of its reactivity, whereas an element soluble in sea water and reactive chemically probably is essential. We look then to the sea for our standards, in order to classify the elements, remembering that some primitive organisms have the ability to concentrate certain trace elements many thousands of times.

Man in his infinite wisdom has chosen to dig up from deposits in the earth a number of metals which occur in traces elsewhere. By fashioning myriads of articles from these metals he has constructed an ever more complicated civilization. New metals, and new applications of common ones, are developed annually. The price of his civilization is an increasing environmental exposure to amounts of metals to which man has been little adapted in the past. In some cases, he has not had time to develop mechanisms for rejecting or excreting these metals, to which he was minimally exposed only 4500 years ago. Therefore, for a half of one percent of his evolutionary history, which has lasted a million years or more, his body has been faced with new and subtle hazards to overcome, and for less than a hundredth of one percent of his history, these hazards have been evermore subtly but rapidly increasing.

We are not concerned with modern exposures to biologically inert elements, even though they are relatively new. We are concerned with the elements which accumulate in living material, and which are biologically active and toxic. As man is the chief concern of this book, we must look at those elements which accumulate in the body of man, and try to discover whether or not they cause human diseases.

The survival and health of the human species depends upon reproduction. Any substance which is toxic to reproduction adversely affects the health of the species. Substances which cause sterility or fetal deaths affect only that segment of the population exposed to them during the child-bearing period. Less toxic substances causing genetic abnormalities which are so severe as to prevent survival to maturity—and the childbearing age—do not affect the health of the species other than the individuals exposed to them. Still less toxic substances, which produce genetic abnormalities of lesser seriousness and allow survival to the reproductive age, affect the health of the race, for they become hereditary disorders which are not naturally eliminated from the race. Much less toxic and more subtle poisons, not affecting reproduction, not causing hereditary and congenital diseases, and not manifesting themselves until advancing age, do not affect the health of the race by interfering with reproduction but merely shorten life span. It is with these last that we are concerned.

It is not within my province to discuss the effects of radioactivity, which, according to dose, can cause all of the results here described. Our problem is with the toxic trace elements. It does not take much imagination or knowledge of industrial practices to discover which trace elements are mined and distributed throughout the world in food, water and air. It takes much detailed work to discover which are in the body of modern man, and which accumulate with age.

In atomic order, industrially based and wide-spread contaminants are beryllium, silver, cadmium, tin, antimony, tellurium, barium, gold, mercury, lead and bismuth. Exposures to other metals occur in workers and miners, but not the general public.

Industrial metals which are non-toxic in ordinary amounts are silver, barium, gold and bismuth.

The metals whose soluble salts in traces taken by mouth shorten life span of animals are cadmium, tin, antimony, tellurium, mercury and lead. Present exposures are by-products of civilization. I will devote separate chapters to cadmium and lead, as these metals induce specific disorders. In table VII-1 are their intakes, outputs, balances and target organs where they accumulate.

It is not surprising that the elements which accumulate in mammalian tissues with age during continuous exposures are toxic and those which do not accumulate are not, as a rule. What is surprising is that the toxic, cumulative elements are generally confined to the metals on the right of the Periodic Table. If you will look at the Table, you will see the positions of those which subtly shorten life spans of rats and mice: gallium, germanium, cadmium, indium, tin, antimony, tellurium (a non-metal), mercury, and lead. (We have not tested the very

TABLE VII-1

Estimated daily balances of abnormal trace elements

	Cadmium μg	Germanium mg	Tin mg
Intake			
Food	200	1.5	4.0 (1-40)
Water	12	?	0 (0-0.03)
Air	1	+	0.003 (0-0.007)
Total	213	1.5+	4.003 (1-40)
% absorption	25	?	2
Output			
Urine	50	1.4	0.023 (0-0.04)
Feces	160	0.1	3.98 (1-40)
Sweat, hair etc.	+	+	0.55
Total	210	1.5+	4.0
Retained in body	3	+	0.0055
Organs retained in	kidney liver	spleen	heart
Total body content, mg	38	20?	5.8

showing recondite toxicity.

Antimony μg	Tellurium μg	Mercury μg	Lead μg
< 100	110	10 (5-20)	260 15-440
48	0	0	20 (0-100)
0.017	0	0-50 (20-760)	20 (2-126)
150	110	10-60	300 (102-666)
Poor	20-50	5-10	5-10
< 70	53	0-35	30 (10-53)
70	57	10	175 (60-400)
1.1	?	0.9	88 (29-202)
< 50	110	10-60	293
+	0.4	+	7 (3-11)
spleen liver kidney	bone	kidney liver	bone
5.9	7	13	121

heavy metals, other than lead, because human exposures are very low.) There is an undisclosed natural law hidden here.

Tin: Civilization took its first great step 3500 years ago with the discovery that copper and tin made bronze, the first durable alloy. Therefore, tin is an ideal metal to suspect of causing disease, for it is scarce on the earth. It has been contaminating preserved foods since the Civil War, accumulating in the human heart and intestine with age (and in rat and mouse tissues), and is subtly toxic in that rat and mouse lives are shortened by 5 ppm in drinking water. However, no specific diseases or microscopic changes are produced.

We can suspect tin as causal in any new disease of the last century which has been increasing in incidence with the increased use of canned foods, and decreasing in incidence with the practice of lacquering the inside of tin cans and the substitution of frozen foods and plastic containers. However, we have no grounds for other than suspicion. Many imported foods are in cans lined with raw tin, as are some domestic baby foods (!); they can be readily identified by looking at the empty can, for the lacquer is a light tan color. The longer a food is stored in tin, the more tin is absorbed by the food. Canned fruit juices are especially high in tin. Tin is scarce on the earth's crust, is inconsistently found in plants, is a poor catalyst, is not present in the newborn infant nor in many wild animals (animals in Vermont are a notable exception) and increases with age and exposure. As much or more tin is ingested in the American diet as chromium, manganese, cobalt, copper and molybdenum, essential elements. So accustomed is the housewife to the taste of tin in canned asparagus, that tin is often added to asparagus preserved in glass to make it "palatable."

If we wish to avoid tin, we avoid tinned foods and stannous fluoride in toothpaste. However, only people who are ultra-cautious and ultra-suspicious of food additives will worry about it, for tin has been of enormous benefit to the human race and has not been shown to cause any chronic human disease. Until research has disclosed specific disorders from tin, it can be largely neglected. Exposures in food have declined about 80 percent during the past 25 years.

Antimony: Little is known about antimony in human metabolism, although it shortened the life spans of rats and mice considerably. Presumably most of the small amount, 5.8 milligrams, in the human body came from glazes on pottery and cooking utensils. Antimony is found in red blood cells and serum, in the air in small amounts, in hair and probably in sweat. It accumulates in lung from airborne sources. This element is unlikely to be responsible for any chronic disease, although it is undoubtedly toxic in larger amounts, causing heart disease in rats.

Mercury: Most of the mercury in the body, 13 mg, comes from the food and air: it is retained in the kidneys and liver, as is cadmium; it is soluble in body fluids and does not accumulate in the lung. With the increased use of mercury compounds in interior paints and in seeds to prevent the growth of molds, exposures are increasing. Because overt mercury poisoning injures the kidneys, this increasing exposure may represent a potential hazard, but present levels are probably completely tolerable without minor injury. Mercury will vaporize from paints in a closed room. Dental amalgam is probably a minor contributor to the amount of mercury taken into the stomach. There is more mercury in the air of American than European cities, from burning coal. It is found in all living things.

Tellurium: A non-metal, tellurium probably enters the human environment largely from heavy metal ores; copper, silver, lead and gold occur naturally as metal tellurides. Tellurium is used industrially in alloys, in rubber, and in many non-ferrous metals to change their characteristics. Therefore, exposures are wide-spread. Because tellurium is not ubiquitous in vegetation and in wild animals, it is a product of civilization and should be treated as such.

Rats and mice exposed to soluble tellurium have shown shortened life spans indicating innate toxicity, but we have been unable to find specific causes of death.

One source of dietary tellurium is tin cans. When a can is opened, some 8 mg of metal filings are deposited in its contents, which are presumably swallowed. (This can be demonstrated by taking a clean empty can opened at one end, inverting it over a sheet of paper, opening it with a can opener and collecting the shavings with a magnet.) The metal shavings contained 0.1 percent tellurium, or 8 micrograms per can. To avoid this source of contamination and irritation, an opener which uncrimps—rather than cuts—cans has been developed but is not yet widely used.

There are at least seven trace elements found in fair-sized amounts in the human body which have shown no toxicity at small exposures. (Table VII-2) Three of them, bismuth, lithium and silver are present in small amounts, less than three milligrams, and are not expected to take part or interfere in biological processes at those levels. The other four are found in larger amounts but are inert, as far as can be discovered: zirconium, niobium, titanium, and barium. There is a tendency for barium to accumulate in bone and titanium in lung, the latter from airborne sources.

Several other metals are found in the body in small amounts: gold ($<$ 10 mg), cesium (1.6 mg), uranium 0.1 mg, radium, minute traces. Although the last two are radioactive, they probably do not cause minor effects at these low doses.

Many if not all other elements could probably be found in traces in human tissues if sensitive enough techniques were developed, but at such concentrations it would strain the imagination to believe that they could possibly cause adverse or beneficial effects on health. We can safely neglect them from our major thesis until one or more of them becomes industrially useful and is introduced in large amounts in food, water or air. There are 13 rare earths and 13 metals in this category. One, thallium, is very toxic and developed a poor reputation from its use as a depilatory. The normal intake is but 2 µg per day in foods, where it is ubiquitous, but one can hardly believe that 2 µg can cause appreciable damage. Thallium in foods cannot be avoided, as it is naturally present.

TABLE VII-2

Estimated daily balances of non-essential trace metals without toxicity

	Barium	Zirconium	Niobium	Bismuth	Lithium	Silver	Titanium
	μg	mg	μg	μg	mg	μg	μg
Intake							
Food	645	3.5	600	0	2.0	70	850
Water	80	0.65	20	20	+	?	2.1
Air	25	+	0	0.01	?	?	0.7
Total	750	4.2	620	20	2.0+	70	853
% absorption	1-15	0.01	40-60	8	90	poor	1-2
Output							
Urine	50	0.14	360	1.6	0.8	9	330
Feces	615	4.0	260	18	1.2	60	522
Sweat, hair etc.	85	+	0.3+	?	+	1	1
Total	750	4.2	620	20	2.0+	70	853
Retained in body	L	0?	0?	0?	0	+	L
Total body content, mg	22	420	120	0.2	2.2	0.8	9

L = retained in lung from airborne metal.

Note: None of these metals is innately toxic to mice or rats at low lifetime doses. Zirconium and lithium are given in milligram quantities, the remainder in micrograms.

TABLE VII-3

Estimated daily balances of potentially toxic elements present in very small amounts.

	Beryllium μg	Thallium μg	Thorium* μg	Uranium* μg	Polonium* $\mu g \times 2 \times 10^{-10}$	Radium* $\mu g \times 10^{-6}$
Intake						
Food	12	2	3	1.3	3.2	2.1
Water	1	0	0.05	0.6	?	0.2
Air	0.01	0.05	?	0.007	0.007	0
Total	13	2	3	1.9	3.2	2.3
% absorption	0.01	95	3	5	10	0.5
Output						
Urine	1.3	0.3	0.1	0.05-0.5	0.011	0.08
Feces	11	1.2	3	1.4-1.8	3.2	2.2
Sweat, hair, etc.	+	+	?	0.0002	+	?
Total	13	1.5+	3	1.8+	3.2	2.3
Retained in body	L	+	+	0.033	L,K,B	B
Amount in body, μg	36	?	?	90	?	0.000031

L = lung B = bone K = Kidney

*Radioactive. 2×10^{-10} is 0.00000000002 μg. Natural uranium has little radioactivity.

Note: These elements all occur naturally in the environment; amounts can be increased by special exposures. Cigarette smokers have more polonium than non-smokers; beryllium workers retain it in lungs, where it causes disease.

In table VII-3 are shown six potentially toxic elements natural to the environment; four of these are radioactive and all accumulate to some extent in the body. Beryllium disease occurs in workers exposed to dusts, and atmospheric beryllium is increasing from industrial smokes and rocket exhausts. Thorium when injected as a contrast medium for X-ray photography stays in the tissues for life. Uranium in large doses causes kidney disease. Polonium in tobacco smoke has been proposed as a cause of lung cancer. Swallowed radium has caused bony changes and in some cases tumors (the "radium girls" who painted watch and clock dials, pointing their brushes between their lips, swallowed large amounts; some of them are still living after 50 years), but the doses employed bear no resemblance to natural exposures, except possibly in the case of polonium, when smokers get 2 μg extra in lungs daily. We have no reason to believe that natural exposures do any harm, for man has lived with them throughout his evolutionary history.

There are 41 trace elements in the environment found in the human body in amounts large enough to be readily detected. Probably 20 of them can be neglected as causing biological effects, good or bad. Seven are products of civilization and are demonstrably toxic. Thirteen are known or suspected to be essential for mammals and one for plants. The remainder, insofar as can be determined, are present in mammals in such small amounts as to be virtually inert.

We must remember, when we speak of toxicity, that a great many elements are toxic in forms other than those present in biological systems. Elemental fluorine, chlorine and bromine gases and crystalline iodine are extremely toxic in decreasing order (fluorine gas will etch glass), whereas sodium fluoride, bromide and iodide have little toxicity and sodium chloride (salt) forms the electrolyte medium in which all animal life except insect, exists. Elemental arsenic can be, and is, eaten, whereas arsenic trioxide is a poison. Elemental selenium has little toxicity, but its oxides and acids are highly toxic. Mercury metal can be swallowed, but mercuric chloride is a deadly poison. Manganese and copper in one valence state are essential

[handwritten margin notes: 41 trace elements in man detectable]

for life but potassium permanganate and copper sulfate are toxic. Most metals in gaseous form are very toxic when inhaled, especially where combined with hydrogen or carbonyl groups. Therefore, toxicity depends on two factors, the nature of the element itself and the form or compound in which it occurs. In this book, we consider each element in the forms in which it occurs in Nature. Industrial toxicity is a special subject which we have not covered, except as it applies to the population at large.

CHAPTER VIII

Cadmium, High Blood Pressure and Water

High blood pressure, or hypertension, is one of the most common chronic diseases in the United States and other civilized countries and is one of the most uncommon in some, but not all, primitive areas. It appears to be linked to environment but not race in a mysterious way, and is not a natural variant in all populations. It is also linked to age, although children can have it. Its severity is increased by emotional tension, although it itself produces emotional tension and anxiety. Some fifteen million Americans—or some five million—or some 40 percent of the population over age 40—have it, depending on who does the reporting—Heart Associations for fund raising, conservative medical men interested in lessening public anxiety, insurance companies out to make money. Deaths directly or indirectly caused by hypertension have halved during a decade and are falling further, owing to the widespread use of new drugs.

High blood pressure causes three changes in the body: enlargement of the heart, disease of the arteries of the kidney, and an increase in the rate of progression of hardening of the arteries. These changes can lead to heart failure, kidney disease, accidents from coronary occlusion, or heart attacks, and brain damage, or strokes, either due to hemorrhage or thrombosis. Therefore, if one has atherosclerosis (as from chromium deficiency) and high blood pressure, one is more apt to have, and die of, a heart attack or a stroke of apoplexy than if one has only one of these diseases. High blood pressure is thus a major contributing factor in serious accidents of the arteries.

People who get kidney diseases may get high blood pressure, but not all of them do. Animals given kidney diseases may get high blood pressure, but not all of them do. Some additional factor seems to be necessary.

We have exposed rats for their lifetimes to small amounts, usually 5 ppm, of soluble metals in drinking water to find out whether we can reproduce in them human diseases. We have given 100 or more rats, half male and half female, vanadium, chromium, nickel, germanium, arsenic, selenium, zirconium, niobium, molybdenum, cadmium, tin, antimony, tellurium and lead. Only in those given cadmium was the full picture of human hypertension developed, with large hearts, changes in the blood vessels of the kidneys, high blood pressure, and an increase in atherosclerosis. Cadmium accumulates in both the human and the rat kidney, arteries, and liver, where it interferes with certain enzyme systems requiring zinc. Cadmium has more of an affinity for certain kidney tissues than does zinc, therefore displacing zinc and changing zinc-dependent reactions. It is also bound by blood vessels.

If we examine the ratio of zinc to cadmium in the American human kidney by weight, we find about 16 milligrams of zinc and over 11 milligrams of cadmium, or a ratio of 1.5. If we look at the primitive African kidney, the ratio is about 6. In the kidneys of beef the ratio is about 40 and of pork 72. In the laboratory rat on a special low cadmium diet the ratio is 464-500 and in the mouse 451. In wild deer it was 23-70, in a coyote 54 and in a dog 24. In laboratory rats fed cadmium and having hypertension the average was 1.7 to less than 1.0. In people dying of hypertension the ratio was 1.4 to less than 1.0. Therefore, in mammals, there was much more cadmium in relation to zinc in Americans and in rats fed cadmium with hypertension than there was in any other mammal studied. The ratio of zinc to cadmium on the earth's crust is 500-1000.

On these and many other studies, representing hundreds to thousands of analyses, we have come to the conclusion that cadmium in kidneys in relation to zinc is a contributing, if not sometimes the whole, cause of high blood pressure. In areas where the ratio is low the incidence of hypertension is high, and vice versa, based on the kidneys of 400 or more human subjects from around the world.

By injecting a special chelating agent, which binds cadmium more than it does zinc, we can remove some of the cadmium from rats' kidneys, replacing it with zinc. When we do this, we cure the hypertension overnight. If we continue to feed cadmium, hypertension slowly returns in several months, but can be cured again. This agent is under study in human subjects and shows promise.

Why does civilized man accumulate cadmium in his kidneys with age? The natural sources are food, water and air. If cadmium can displace zinc in his body, it is likely that food containing more than usual amounts of cadmium and less than usual amounts of zinc might slowly lead to the accumulation of cadmium. An excess of zinc would prevent accumulation of cadmium, a slight deficiency allow it. In table VIII-1 are shown the amounts of zinc and cadmium in 250 common foods, and their ratios. Also in the table are a few representative foods with relatively large amounts of cadmium.

We can obtain a few hints from the table as to the behavior of cadmium and zinc and their ratios. First of all, there is relatively much cadmium in oils and fats and relatively little in nuts, which are very fatty, whole grains, which contain fats, and legumes. There is much cadmium in oysters, but there are large amounts of zinc to cover it. If we take a ratio of zinc to cadmium of 60, or half the average, we find that most foods have ratios below it. If we take the dividing line at a third the average, or 40, we find that whole grains, legumes, root

TABLE VIII-1

Average concentrations of zinc and cadmium and their ratios in foods, ppm

Food	Zinc	Cadmium	Zinc/Cadmium
Seafood	17.5	0.79	22
Oysters	1280.0	3.40	378
Clams	27.8	0.58	48
Canned anchovies	17.7	5.39	3
Smoked kippers	20.3	1.28	16
Meats	30.6	0.88	35
Lamb chop	53.3	3.49	15
Chicken	29.0	2.0	15
Dairy Products	1.0	0.27	4
Butter	1.8	0.56	3
Cereals and Grains	17.7	0.16	111
Gluten	48.5	0.51	97
Polished rice	1.6	0.06	27
Vegetables			
Legumes	10.7	0.03	357
Roots	3.4	0.07	49
Leaves and fruits	1.7	0.13	13
Oils and Fats	8.4	0.75	11
Olive oil	2.8	1.22	2
Margarine	1.6	0.8	2
Nuts	34.2	0.05	684
Fruits	0.5	0.04	13
Beverages	0.9	0.07	13
Average ratio			119
Whole human diet #1	6.1	0.19	32
diet #2	5.0	0.12	42
Purina Laboratory chow	58.1	0.63	92
Blue Spruce Farms rat chow	31.7	0.02	158

vegetables and nuts are above it, and that seafood, meats, dairy products, leafy vegetables, oils and fats, fruit and beverages have enough zinc to cover the cadmium they contain. We do not know what a "safe" ratio is, that is, one which would prevent accumulation of cadmium; we can only guess.

We can also suspect contamination by cadmium in the processing of some of these foods low in zinc: canned anchovies, smoked herring, chicken, olive oil, margarine, canned fruits and beverages. Cadmium is a constant contaminant, up to 1.5 percent, of the zinc used in galvanizing; as this protective coating is cheap, it may be used in food processing in bulk. Milk cans are galvanized, for example. Foods, especially acid ones, in contact with galvanized surfaces will absorb cadmium. In fact, acute cadmium poisoning has occurred several times in acid drinks left in galvanized pails or ice trays.

Superphosphate fertilizers contain much cadmium. Derived from large deposits of prehistoric fish teeth and bones, principally in "Bone" Valley, Florida, they have absorbed and stored cadmium from sea water which now appears in fertilizers to be spread over the land. One Yugoslav deposit has so much cadmium that it can be partly removed and sold as a by-product of the preparation of superphosphate fertilizer. At first we believed that superphosphates might be a source of cadmium in vegetable foods. When we grew grains on an abandoned pasture far from sources of environmental contamination, we found little enrichment of cadmium in those heavily fertilized (table VIII-2). Presumably the presence of much zinc prevented contamination with cadmium.

From the ratios, the table shows little change in grains, more zinc in legumes and a higher ratio in root vegetables, no change in leafy vegetables, some decrease in the ratio in fruity vegetables and no change in vinous vegetables, even in the presence of 9 ppm cadmium. In fact, the average ratio went up. Thus we can probably exclude these valuable adjuncts to food production, which are full of trace elements including fluorides, as sources of food cadmium.

We ask ourselves if the refining of foods could remove zinc (as shown in Chapter VI) and not remove cadmium. That this happens is elegantly shown in table VIII-3. The refining of

TABLE VIII-2

Zinc and cadmium and their ratios in vegetables and grains grown on soil fertilized with superphosphate, ppm.

Food	Control			Superphosphate		
	Zinc	Cadmium	Zinc/Cadmium	Zinc	Cadmium	Zinc/Cadmium
Grains	17.1	0.15	114	22.2	0.16	139
Legumes	0.6	0.03	20	4.3	0.05	86
Roots	3.1	0.10	31	2.6	0.02	129
Leaves	2.4	0.05	48	3.5	0.09	39
Fruits	1.8	0.05	36	1.8	0.09	20
Vines	1.6	0.01	155	1.5	0.01	147
Average			67			93
Soil, grains	71.0	0.27	263			
Soil, vegetables	62.0	0.20	310			
Superphosphates, 20%				129.4	8.97	14

TABLE VIII-3

Zinc and cadmium and their ratios in major carbohydrate and sources of calories and their fractions, ppm

Food	Zinc	Cadmium	Zinc/Cadmium
Grain			
Wheat	31.5	0.26	121
Farina	6.6	0.27	24
Patent Flour	6.4	0.38	17
First Clear Flour	15.9	0.26	61
Low Grade Flour	33.6	0.26	129
Red Dog	105.3	0.28	376
Shorts	106.0	0.92	115
Bran	100.2	0.88	113
Germ	133.4	1.11	120
Bread, whole wheat	5.3	0.15	35
Bread, white	1.2	0.22	5
Rye, whole	32.5	0.01	325
Wheat, gluten	48.5	0.51	97
Rye, oil	6.9	0.07	98
Oats	33.5	0.19	177
Barley	4.4	0.09	49
Buckwheat	25.5	0.27	95
Corn	3.8	0.12	32
Rice, American, polished	0.6	0.04	15
Rice, Japanese, polished	1.6	0.06	27
Sugars			
Sugar Cane	1.0	0.04	25
Granulated	1.3	0.06	22
Lump	0.2	0.27	<1
Molasses, household	2.9	0.56	3
" , refining	3.3	0.83	4
" , black strap	8.3	0.86	10
Honey, refined	1.0	0.74	1
Fats			
Olive oil	2.8	1.22	2
Lard	0.5	0.05	10
Margarine	1.6	0.80	2
Butter	1.8	0.56	3

whole wheat, with a zinc:cadmium ratio of 121 and plenty of zinc, results in flour with a ratio of 17 and little zinc. The ratio progressively rises in the middlings, or the parts fed to cattle, just as does the zinc. The point to be noticed, however, is that cadmium is held largely in the endosperm of wheat, whereas zinc is held largely in the bran and germ. No increase of cadmium was found in low grade flour or red dog, whereas zinc increased 15 times. Whole wheat bread had an acceptable ratio of 35, white bread an inacceptable one of 5. From these analyses we can conclude that human beings eating white bread and white flour products do not get enough zinc to cover their intake of cadmium, or to put it another way, they take in relatively more cadmium than they should for the amount of zinc they get.

You will also see that the whole grains rye, oats, barley and buckwheat do well with their ratios of zinc to cadmium, but that polished rice does not. Japanese workers have found that cadmium is similarly enriched in polished rice, 79-89 percent of it remaining after polishing 36-40 percent of the grain away, this discarded fraction containing most of the zinc and other elements.

About 90 percent of the cadmium in rice is firmly combined with the protein glutelin, and about 90 percent of the cadmium in wheat is combined with gluten. This is probably so in other grains, cadmium bonding to protein as it does in kidney. The Japanese, with their usual thoroughness, analyzed a thousand samples of rice and found marked geographical variations in cadmium content. They showed that 100 grams of Japanese rice, or less than they eat a day, would provide 2-189 micrograms of cadmium, American rice 6-137 μg (Mississippi being the highest and Louisiana and Arkansas rice the lowest), Formosan rice 31-70 μg, Spanish 49 μg, and Egyptian, Burmese and Thai rice 13-19 μg. Many Orientals eat unpolished rice, but not the Japanese.

The table also shows the low ratios in sugar and molasses, compared to sugar cane, in processed honey and in fats and oils. The high values in olive oil, margarine and butter may come from contamination during the processing. Therefore, we have ample reason to pinpoint the relatively high intake of cadmium as compared to the low intake of zinc to refining and processing the major sources of calories in the United States and other wheat-eating and rice-eating countries.

Another possible source of cadmium is in common beverages. This source is explored in Table VIII-4. The ratios are low in coffee, but there is little zinc. A liter of coffee a day, 6-7 cups, would supply only 6 μg cadmium; if coffee 97% caffein-free were used (although who would want to drink that much?) 4-15 μg would be drunk. Tea supplies 6-10 μg cadmium to serious tea drinkers taking a liter a day, and the ratios are higher, for there is more zinc than in coffee. Cocoa, prepared orange juice and grape juice have acceptable ratios. Therefore, we can exclude these as major sources.

Alcoholic drinks consumed in quantities have more cadmium than coffee and tea. A liter of wine contained 49-80 μg and a liter of vermouth contained 50-170 μg. For alcoholics, a liter of gin had 20 μg, of Bourbon whiskey 100 μg, of Scotch whisky 50 μg and of brandy 90 μg. As the total daily intake of cadmium is only 200 μg, these amounts represent sizeable increments.

It is very likely that cadmium in drinking water and other fluids is more readily absorbed from the intestinal tract than cadmium bound to foods. Therefore, we must examine tap water and other thirst quenchers for cadmium. Table VIII-5 shows these analyses. Eight natural soft waters had 1.5 μg per liter, whereas water piped through copper pipes had 10.2 μg, through black polyethylene pipes had 13 μg, and through galvanized iron pipes had 16.3 μg per liter. All of these values exceed the limits of cadmium in potable water allowed by the U.S. Public Health Service, or 10 μg per liter. It is clear from this table that soft water can take up sizeable amounts of cadmium from any common type of pipe, cast iron, plastic, copper or zinc lined, except brass. Also large amounts of zinc are sometimes dissolved, especially from copper and galvanized iron pipes by soft water, up to 2 ppm. If all of the cadmium in

TABLE VIII-4

Zinc and cadmium and their ratios in tea and coffee, soft and alcoholic drinks, ppm

Beverage	Zinc	Cadmium	Zinc/Cadmium
Coffee			
Infusion	0.03	0.006	5
, dried	0.27	0.37	< 1
Instant	0.10	0.006	16
, dried	2.63	2.27	1
, ground	3.10	0.32	10
Average			6
Caffine-free			
Infusion 1	0.02	0.009	2
2	0.04	0.015	3
3	0.12	0.004	30
4	0.13	0.011	12
Dry 1	5.31	1.39	4
2	5.68	0.67	8
3	3.64	2.05	2
4	1.39	0.65	2
Average			8
Tea			
Infusion 1	0.33	0.01	33
2	0.24	0.006	40
dried	0.44	0.56	< 1
Black, leaves	34.36	1.61	21
Green, leaves, Japanese	36.30	2.50	11
infusion	0.04	0.001	40
Instant	0.20	0.008	25
dry	8.00	1.38	6
Average			29
Cocoa, dry	48.65	0.67	73
Orange juice	0.65	0.01	65
Grape juice	3.11	0.07	44
Wines	1.41	0.085	17
Beers	0.33	0.01	33
Whiskey and gin	0.09	0.065	1

these waters were absorbed into the body and retained in the kidney, in a year's time 3.7 milligrams and in three year's time 11.2 mg would have accumulated, or the total content of the American kidney. In 10 years time, the water drinker would probably have high blood pressure.

We do not know exactly where the cadmium in copper and plastic pipes comes from. Metallic plasticizers are used to harden black polyethylene and we know that these pipes give up both cadmium and lead to soft water. An installation of 210 feet of black polyethylene from a forest spring to our animal laboratory gave up so much lead and cadmium to soft water that we had to install a water purifier. As to copper pipes, we have found 3-57 ppm cadmium in the pipe itself; some must come from fittings or solder. In galvanized iron pipes there is 140-400 ppm cadmium.

We also found 15 μg in a liter of a widely advertized cola drink, with only 2.5 μg in ginger ale, 0.5 μg in birch beer and 2.0 μg in another cola. Only the one cola exceeded official limits of potable water of cadmium. Therefore, in view of the avidity of cadmium for kidney, adequate sources for accumulation are in most soft piped waters and a popular bottled drink.

Soft water corrodes metal pipes, especially when it is acid. Any water, soft or hard, with a pH of less than 8.0 can be corrosive, below 7.0 it is always corrosive (a pH of 7.0 is neutral). It becomes slightly acid from carbon dioxide dissolved in it from air, and from decaying organisms occurring naturally in reservoirs. Surface water from lakes and rivers is naturally soft, for it comes largely from rain which does not long remain in contact with minerals in the earth. Soft water will dissolve copper from pipes, as clearly shown in table VIII-5. Any metal dissolved from water pipes goes down into the stomach if the water is drunk. When the metal is iron or copper, it does not matter, but when it is cadmium, we must be concerned. So corrosive is soft water that my plumber friends tell me that almost all galvanized pipes in old houses in our town have had to be replaced with copper. Even copper pipes can spring leaks

TABLE VIII-5

Zinc, copper and cadmium in water, ppb

Water	Zinc	Copper	Cadmium	Zinc/Cadmium	Remarks
Natural waters					
Connecticut River, Vt.	5	—	14.6*	<1	polluted
Brook, Vermont	14	1	0.5	28	clear
Brook, Vermont	3.5	—	0.6	6	clear
Spring, N. Hampshire	177	—	2.5	70	clear
Sea water, Caribbean	24	—	0.3	80	clear
Municipal water, town, Vt.					
Reservoir, inlet, brook	3.5	16	2.1	2	
Spillway	3.5	55	2.5	1	
Main, town	—	150	14-21*	–	cast iron pipe
Tap, hospital, cold running	160	170	8.3	19)	galvanized
, stagnant	1830	730	15-77*	122-24)	and
, hot, running	1800	440	21*	86)	copper pipes
Tap, institution, cold, running			1.0		brass pipes
Tap, Bridgeport, Conn.	—	185	14.0*	–	galvanized pipes
White Plains, N.Y.	—	540	14.0*	–	galvanized pipes
Bangor, Pa. municipal	—	—	0	–	

Lake, N.H.	3.5	—	1.1	3	
, tap	20	—	14.0*	1	plastic pipe
Spring, Vermont	13	3	2.2	6	
, tap	660	1460	3.8	174	galvanized pipes
Well, Vt., tap, hard	100	36	1.1	90	galvanized pipes
Well, Vt., tap, soft	2160	279	3.5	616	galvanized pipes
Spring, Vt. tap	770	1240	8.3	93	copper pipes
Well, Vt. artesian tap	219	4	12.0*	18	copper and plastic pipes
Spring, S.C.	18	—	8.0	2	galvanized pipes
Snow, melted 60' from street	1380	—	1.5	920	moderate traffic
, forest hilltop, 300'					
from road	45	—	0.35	129	little traffic
, protected	8	—	0.38	21	no traffic
Summary, natural waters	30	2	0	49	
Summary, copper pipes	495	907	10.2*	52	
galvanized pipes	844	—	16.3*	—	
brass pipes	—	—	1.0	9	
plastic pipes	120	—	13.0*		

*Cadmium concentrations exceeds allowable limits for drinking water, U.S.P.H.S. standards.

after a long time, as I can attest personally. The dissolved copper leaves a blue-green stain in bathroom fixtures.

Hard water contains calcium and magnesium bicarbonates which precipitate on metal pipes, laying down a hard coating which prevents corrosion, unless the water is acid—which is unusual. If the water is very hard, scale collects to the point that the pipes become plugged. We can expect little cadmium, iron, zinc or copper in hard piped water. In some areas, water is so hard that a precipitate forms when it is added to whiskey; soap will not lather and bathtubs have rings around them. You can tell if your water is hard or soft by whether or not soap lathers or it is difficult to wash off soap from your hands. Most deep well waters are hard.

Since World War II a major revolution has taken place in plumbing. Hourly wages have sky-rocketed to the point where plumbers make more money than most scientists. Galvanized iron pipes must be threaded, jointed, angled and fitted, which takes much time. Copper pipes can be bent and soldered. Although copper of itself is more expensive, it is cheaper to install, and all new houses have copper pipes and fittings. The day of the steam radiator is over.

In the United States and in most large countries water varies in hardness according to geographical area and whether or not surface or well water is used in municipalities. In Japan, all water comes from rivers, and is soft, just as is the water on the eastern seaboard, the slopes west of the Sierras and the Gulf and Great Lakes shores, where there is much rainfall. The Central and Mountain States in general have hard water.

In 1957, I met Jun Kobayashi in Tokyo, who showed me data indicating that there was a positive relationship between the death rates from apoplexy (cerebral hemorrhage) and the sulfate-bicarbonate ratio of river water in Japan, according to geographical area. This ratio is an index of the acidity of water; cerebral hemorrhage, the first cause of death in Japan, is almost always the result of high blood pressure. I was also shown a relationship by prefecture (state) between the acidity of water and the prevalence of high blood pressure, according to insurance statistics. When I returned to the States, I tested this idea statistically in Japan and found it significant.

At that time, the U.S. Geological Survey had compiled data on some 25 qualities of municipal water in over 1300 cities and towns in this country, made in 1949-51. Most of these data were on bulk elements. The Department of Health, Education and Welfare had for the first time a detailed list of death rates from major diseases by age and state for 1949-51. So it seemed rewarding to test what quality of water there was available by state and death rates from major diseases by state. The only quality of city water that had been worked out by state was hardness. Death rates from cardiovascular diseases varied widely from state to state.

After a tortuous self-taught course in statistics and reams of paper, a relationship was found. It was inverse, or negative. The softer the water, the larger was the statewise death rate from high blood pressure and hardening of the arteries—coronary occlusion, stroke—but for no other major disease including cancer. Thus was opened the Pandora's box of water quality and heart-artery diseases. As I said before, a direct relationship existed in Japan between the degree of acidity of river water and apoplexy, all Japanese waters being soft. Therefore, it was obvious that something other than hardness of water was involved.

Using a Model T computer, we also found that death rates by city from coronary occlusion—atherosclerotic heart disease—were inversely related to hardness and to the amount of calcium, magnesium and a few other elements in municipal water. From this, others have assumed that soft water and heart disease were related, forgetting the Japanese experience with strokes and acidity.

These data were reported shortly after publication at an international meeting in Prague. J.N. Morris, who had seen my abstract to discuss my paper, had time enough to confirm the findings in England in a very preliminary way. Later he and his colleagues confirmed it in toto. This was the only time I have reported a new idea at a meeting and had it confirmed on the spot.

There was considerable publicity. The water-softening industry became alarmed. They apparently hired five well-known experts to go into the problem, none of whom was intimately conversant with all of its ramifications. Their report was published in an obscure journal; it stated that the association was "spurious," and questioning people who wished to buy water softeners were handed a reprint of their report. Morris and my answers to this report were not allowed to be published. Thus do vested interests buy scientific opinion, pervert those scientists who can be perverted, and delude the public when profits are threatened. No mention was made by us about softened water. I hope that these so-called "experts" were embarrassed when our findings were confirmed in Sweden, Ontario and Finland, but perhaps they were not.

To this point, there is some quality of water related to hardness or softness and which is not hardness or its salts, but might be acidity, which affects high blood pressure and strokes in Japan (but not heart attacks), high blood pressure, heart attacks and strokes in the United States and England, heart diseases in Sweden, heart attacks with sudden death in Ontario and heart attacks in Finland. Other causes of death are not affected. I have already outlined the answer.

The one disease which fits what is known is hypertension. High blood pressure increases the rate of progression of hardening of the arteries, and its influence would account for differing mortality rates. The one trace element found in soft piped water which causes hypertension in animals is cadmium. People with hypertension have a low zinc-cadmium ratio in their kidneys.

Therefore, in soft water areas where there is a tendency to hypertension in the population, there is more atherosclerosis, more strokes and more heart attacks than in hard water areas. There is enough more cadmium in water to account for these differences. The differences, however, are not caused by the soft water directly, but by the tendency for soft water to be acid, and therefore corrosive. Thus, Kobayashi's original hypothesis, that it is the acidity that counts, is probably the correct one.

I must emphasize that it is not the water cadmium which primarily causes the hypertension, for this disease is prevalent in both hard and soft water areas, although more frequent in the latter. It is the water cadmium which causes the difference. The change in zinc-cadmium ratios of food presumably is fairly constant, a result of refining and processing foods, so that a large fraction of the population is exposed to less zinc than needed to cover the cadmium. When water cadmium is also added to food cadmium and low food zinc, differences become significant.

One other source of cadmium is air, the metal coming from industrial smokes and the burning of coal and petroleum products (see levels in snow, contaminated by auto-exhausts, in the table). Carroll found a high correlation of air-borne cadmium and hypertensive deaths in 28 American cities. The higher concentrations, however, were only 0.02 μg per cubic meter of air; we breath daily about 20 cubic meters of air weighing 52 pounds; at this level the increased cadmium intake is only 0.4 μg, a small amount compared to water. Air-borne cadmium could contaminate the soil, however, in rainfall, although only to the extent of about 4 μg per liter of rain. Cigarette smoke also contains sizeable amounts of cadmium, most of which is absorbed.

Hickey and his colleagues have analyzed death rates from 6 degenerative diseases and the amount of cadmium, vanadium, lead, manganese, tin, zinc, chromium, total dust and benzene soluble matter in air of 26 cities of the United States. There was a positive correlation of cadmium with all heart diseases, and of cadmium plus vanadium with heart diseases. No significant relations were found with any air-borne metal or substance and kidney disease, diabetes, lung cancer, general arteriosclerosis, or hypertensive heart disease, thus pin-pointing cadmium and heart disease as being related.

Therefore, we have another example of a chronic human disease induced by an abnormal trace element accumulating in a target organ, the kidney, which was not meant to be there. The accumulation is the result, as best we can determine, of a) refined foods, b) water pipes and c) contaminated air. Thus has man released deposits of an abnormal metal into his environment, and he is paying the price in terms of his health.

CHAPTER IX

Lead: an Increasing
Potential Hazard

Lead is one of the most abundant pollutants of the human environment and contaminants of the human body. Its principal source today is tetra-ethyl lead, an antiknocking chemical added to gasoline. Over two pounds of lead in the form of tetra-ethyl or tetra-methyl lead per man, woman and child in this country is burned in gasoline yearly; half to three quarters comes out of the tail pipes of automobiles, to pollute the air, food plants and water and contaminate human beings breathing the air, eating the food and drinking the water. Among the benefits of this pollution is the ability to make jack-rabbit starts and getaways with high powered cars without engine knock.

Natural lead occurs in low concentrations (10 ppm) in the earth's crust; only in deposits is it profitable to mine. There has been a slow increase in crustal lead since the earth began. All of the heavy natural radioactive elements, uranium, protactinium, thorium, actinium, radium, franconium, radon, astatine, and polonium eventually disintegrate to lead, and it is possible to measure the age of the earth by taking advantage of this process—it is 3.25 ± 0.75 billion years, more or less. The increment is small, because these elements are so scarce. Primitive man was exposed to little lead, and his bones, which collect lead under higher exposures, have shown none when analyzed.

Since the age of metals began, when man first learned to extract them from ores and work them, exposures to "industrial" lead have vastly increased. Man was exposed to lead in Asia Minor 4500 years ago as a by-product of silver smelting. Because lead is an insidious and slow acting poison, man continued to use lead for 4300 years without suspecting toxicity. Only for about two centuries has he become aware of some of the extreme toxicological effects of lead and learned to avoid them. In the past fifty years more and more of these effects have been found.

Lead is easily smelted, easily worked, and lends itself readily to alloys. It is virtually indestructible. The highest known exposures of human beings to lead, before the age of gasoline, probably occurred in Ancient Rome. Lead amphorae stored syrups and wines, lead pipes carried water to the houses of the rich (when water was soft, it dissolved lead), lead cosmetics were used by the ladies. There is little doubt that lead poisoning was endemic among those who could afford such luxuries. In fact, Gilfillan believes that lead poisoning resulting in stillbirths, spontaneous abortions and sterility, was responsible for the low birth rate of the upper classes of the Roman Empire which led to the ultimate fall of Rome.

There was little lead found in the bones of Third Century monks, but large amounts occurred in those of the Eleventh to Nineteenth Century. The use of pewter and lead food containers were probably responsible. In more recent times, lead pipes have caused chronic lead poisoning whenever soft surface or rain water flowed through them. Up to the present day, we have seen several cases in Vermont, and it has been widely reported in England. Lead exposures, with the advance of Western Civilization, have come from the sealant of tin cans (since 1824, widely since 1865), pewter, candy wrappings, toothpaste, cigarettes (since 1900), insecticides such as lead arsenate (since 1894), paint (B.C. or thereabouts), an addition to bottled wines (since 1800), kitchen utensils made of lead alloys and a score of other uses of this valuable but toxic metal. In recent years, with the spreading understanding that lead can be toxic and produce symptoms of lead poisoning, various obvious sources of exposure have been lessened or eliminated entirely by government action. Not so gasoline-borne lead.

In 1924 tetra-ethyl lead was first added to gasoline on a trial basis, depending upon ultimate toxicity. Thus was started the most wide-spread pollution of the human environment that has ever occurred for any non-essential element but silver, and for any toxic element. Today volatile lead enters the upper atmosphere, to be deposited in glacial ice on the northern coast of Greenland, where it increases annually. Dated samples of ice reveal little detectable lead deposited before 1940,* but each subsequent decade there is more. Lead is found in the depths of the ocean and on the tops of the Sierras; analysis of its isotopes shows it to be mined lead. In Antarctic ice there is little or no lead increasing annually as there are fewer cars in the Southern than in the Northern Hemisphere. Tree rings grown 100 years ago in elms in a suburban area 100 yards from a road had little lead—from 1860-1930; those grown from 1937 to 1960 showed rapidly increasing amounts consistent with the increasing pollution of air-borne lead. At present, grass and weeds grown near a well-traveled town road have enough lead to poison a cow eating them exclusively. Leaves apparently absorb lead from air and rainfall. There is almost no lead in snow fallen in a mountain forest—and much near a road. Air-borne lead in cities is proportional to the consumption of gasoline. Calculations indicate that of the lead entering the body and retained, nearly half of that accumulating in city dwellers, comes from air. The makers of tetra-ethyl lead point with some pride to the fact that the amount of lead in the air of our largest cities has not increased in ten years. The reason is obvious: no more autos can get into a city saturated with traffic. Mountain air contains about 0.1 microgram per cubic meter, rural air 0.5, cities in

*No lead was detected in glacial ice deposited in 800 B.C. By 1750, there was a small amount of lead resulting from smelting of lead ores in the Northern Hemisphere; this amount slowly increased for the next 190 years as the industrial revolution developed. After 1940 the curve increased sharply upward, reflecting the lead emitted from automobile exhausts. There is little intermixing of upper level air between Northern and Southern Hemispheres.

general 2.4 and areas of dense traffic up to 44.5 micrograms. (A person breathes about 20 cubic meters of air a day; about half the lead in it is retained.) It is quite likely that mountain air a century or two ago contained no detectable lead, and that cities had only a small amount coming from coal and wood smoke.

We have analyzed human tissues from various sections of the world for lead. In Americans sizeable increases according to age were found in aorta, pancreas, spleen, kidney, lung and bone; these increases did not occur in foreigners, except in aorta. No lead was found in infant's bones, but bone lead increased in every decade up to the fifth. Most foreigners seemed to be in a steady state, excreting as much lead as they retained. In Americans, the exposure exceeded the ability to excrete. In ten years, the estimated body content of Americans has risen from 80 milligrams to 120 milligrams, but it was 62 milligrams in Swiss, 63 in Africans, 78 in Middle Easterners and 94 in Orientals, in all cases less than in Americans. Bone lead was half or less that of Americans in adults from Japan, Taiwan, Bangkok, Manila and India. There were wide variations in bone lead; in Americans from 10-435 milligrams; in foreigners from undetected amounts to 80 milligrams.

Whereas it is clear that the increasing pollution of the air with lead, proportional to the number of passenger miles traveled by auto, and the increasing contamination of the human body presents a potential hazard to health, we do not know if the danger is here now or at some time in the near future. We do not know if there is such an entity in the population as sub-clinical toxicity of lead, manifest by shortened life span, lessened immunity to infections, feeling poorly, tired, "run down," subject to nervousness, insomnia, lack of appetite and other vague and ill-defined symptoms from which so many of the urban population suffers. No one with imagination, to my knowledge, has examined such persons for excessive lead in their bodies, by giving a chelating agent which removes lead, and testing the urine for large excesses. We found one such case by chance; there may be others. We do know that such persons can improve remarkably when they take a vacation in another environment, at the seashore or in the mountains away from the city, and it would be worthwhile to examine them with this possibility in mind. We do not know what symptoms are common in traffic policemen and garage workers exposed to excessive lead as well as carbon monoxide. We do not know if some diseases, especially those of the central nervous system, are caused by lead. We recognize overt lead poisoning in children eating lead paint, which can result in mental retardation and kidney disease later in life, in painters, in battery workers, in lead smelters and refiners. But in the broad spectrum between actual poisoning and minimal body burdens of lead, we do not know at what point toxicity, be it ever so subtle, occurs. In the case of most poisons, where a large amount carries obvious damage, smaller amounts cause less damage, and the damage is proportional to the dose. But in the case of lead, the idea that there is no damage done by increasing doses until overt symptoms appear has been vigorously defended by the lead industries, who claim there is no clear and present danger from atmospheric lead, and that present exposures are tolerable.

On January 28, 1969, the Poison Board of the Swedish Government decided to lower the present limit of lead allowed in Swedish gasoline from 0.84 grams of lead per liter to 0.7 grams, by January 1, 1970. This decrease resulted from an exhaustive survey of the problem and would in effect retain present exposures of the population in spite of the increase in gasoline consumption. This was the first time a government had acted on the possible hazards of air-borne lead.

We can turn to experimental animals exposed to lead to learn what conditions might develop in human beings. Unfortunately the results in our hands are equally vague.

1. When tetra-ethyl lead was injected into rats which had learned to climb out of a simple water maze, the rats became extremely nervous and eventually had convulsions. There was no suppression of the rate of learning the maze or memory for the maze's exit, even when the rats were extremely excitable. An analogy in man would mean that no decrement in

performance could be expected from mild poisoning by this lead compound. (So poisonous it is that safety regulations required that the 15 milliliters—half ounce—we received had to be transported in a triply insulated sealed pressure bottle of a liter size in a heavy wooden box by automobile personally from the factory some 600 miles away. This amount is what we would put into five gallons of gasoline.)

2. When we fed chromium deficient rats small amounts of lead in drinking water for their lifetimes and reproduced human (American) tissue concentrations, they grew well but there was a significant mortality in the young due to infections, a continuous mortality at all ages, and a median (half) life span which was, excluding the young dead, 249 days less than the life span of their controls, a decrease of 31.2 percent in males, and 218 days, or 22.4 percent in females. As chromium deficiency with lead excess is frequent in adult Americans, the experimental counterpart of this situation showed obvious but undefined toxicity. There were no definite pathological changes produced. In chromium deficient mice, with human tissue concentrations of lead, males lived 92 days and females 53 days less than their controls, a reduction of life span of 9.6 and 5.5 percent. The greatest cause of death was from infections. When chromium supplemented male rats were fed lead, there was no increase in mortality, although older animals lost much weight and hair, were lethargic and looked unhealthy.

3. Male rats and mice given lead were susceptible to infections. During an epidemic of pneumonia, 78.5 percent of lead- and chromium-fed rats died, compared to 34.6 percent of controls. Of young rats fed lead alone 16.4 percent died of infections; no controls died. Infections caused the deaths of 38.5 percent of mice fed lead compared to 11.4 percent of controls. Thus, lead at present human tissue concentrations appeared to suppress immunity to infectious diseases whether or not chromium was present. No chronic diseases, such as high blood pressure, diabetes, elevated blood cholesterol, tumors or arthritis appeared. One cannot help wonder if this suppression of immunity has a human counterpart.

The trouble with lead is that there is no easy way to detect low or high body burdens, unless toxicity has appeared. Lead in blood and urine are indicative of exposures shortly before the specimens were taken. Lead in hair is not correlated with lead in body, except in chronic poisoning. As 91 percent of lead is in bone, from which it may be mobilized (although how and when is not known), the only way to test for it properly is to analyze a sample of bone marrow. Probably the best method is to inject or ingest a small amount of a chelating agent and measure the amount of lead which comes out in the urine. A quick test for lead is badly needed. There is such a test, but all urban dwellers show it: interference with a blood-forming enzyme.

What can we do to avoid accumulating lead? We can do nothing personally, if we live in a large city and travel by car in heavy traffic. We can keep away from soft water flowing through lead pipes. Most new pipes are copper, although in cities there are often lengths of lead pipe leading from water mains to houses. (Hard water deposits calcium salts on pipes, preventing corrosion.) We can avoid eating vegetables grown near well-traveled highways, if we know their source. As only about ten percent of the lead we eat is absorbed, the amount avoided by these means is small, except in the case of lead pipes. Estimated daily lead balances in man are shown in table IX-1, which indicate the relative amounts absorbed from food and air.

The solution to this problem lies in the removal, or severe restriction, of lead additives to gasoline. The price we pay for jack-rabbit starts and rapid acceleration of our cars is the increasing potential hazard of lead. Only one American oil company at this writing makes high test gasoline without lead; it uses the Houdry process of cracking petroleum which produces a high octane gasoline. Conversion of the rest of the oil industry to this process would add one to three cents to the price of gasoline. Because a high test gasoline competitive with leaded gasoline is widely used, the problem is not unsurmountable.

We could insist that automobile engines be designed (as

TABLE IX-1

Estimated daily balance of lead, and ranges

Intake, μg

Ingested		Absorbed	% Absorbed
Food	260(150-440)	26 (8-62)	5-14
Water	20(0-100)	2 (0-10)	10
Air	20(2-130)	10 (1-80)	30-60
Total	300(152-670)	38 (9-152)	

Output, μg

Feces	175(60-400)
Urine	30(10-53-170)
Sweat	65(28-160)
Hair, nails	23(1-45)
Retained	7(3-12)*
Total	300(102-670)

Note: Virtually all of the lead in food and all of it in water and air comes from contamination in the production, preparation and storage of foods, from contamination of piped water and contamination of air from industrial smokes and especially motor vehicle exhausts, which in turn contaminate the soil. Under severe exposures, half of the lead absorbed can come from air.

*121 mg in 50 years

they are being designed) to obtain increased performance on low octane gasoline without lead. Most European automobiles perform well without high test gasoline—lead additives, for example, were not used in Switzerland until 1947.

We could explore methods for preventing the emission of volatile and particulate lead from exhaust fumes into the atmosphere—this would be difficult. We could develop engines running on other sources of power—electricity and steam—which would avoid pollution of all kinds from cars.

The public must decide whether or not it wishes to pay for increased performance of heavy, powerful automobiles, with an increasing pollution of the air and an increasing body burden of lead which may soon lead to ill-health. Civilized man has removed vast amounts of lead, a toxic metal, from large deposits in discrete areas, and is spreading it thinly over the surface of the earth.

CHAPTER X

Balancing Your Diet to Correct Partitioning of Foods

The major foods eaten by man contain the micronutrients and the bulk elements necessary for their metabolism. This repetitive statement is obvious, from an evolutionary viewpoint, for man would have suffered deficiencies and would probably not have survived if this were not so. Therefore, whole and unprocessed foods are balanced in the sense that they are completely utilized by the body. If this were not so, primitive omnivorous man would have had to select his foods intelligently and balance his diet on the known contents of micronutrients, of which he had no suspicion, or no taste.

Let us say that food A, meat, for example, was deficient in one or more micronutrients—vitamins or trace and bulk elements. Whenever he dined on meat, he would have to go elsewhere to a food, B, which contained the micronutrients for the metabolism of food A, salads for example. Suppose that food B was deficient in another nutrient necessary for its metabolism. In that case, he would have to go to another food which had the nutrient, milk, for example. In this case, to balance his diet and be healthy he would necessarily dine on meat, salad and cheese. Such a selective diet is inconceivable for primitive man, or for any other animal.

The only mammals requiring a balanced diet are laboratory animals, domestic animals and man. No wild animal needs a balanced diet, for it is already balanced in natural foods to which the animal is adapted.

When the first man rendered lard from animal fat or pressed olives for the oil, he fractionated or partitioned a food into two parts. If he had been knowledgeable, he would have assumed that the micronutrients necessary for the metabolism of the oil or fat went with the fat, and the micronutrients necessary for the metabolism of protein stayed with the olive fruit or meat. He might or might not have been correct in that assumption, as we shall see.

For a long time, probably the only common partitioning of foods was the separation of oils and fats from the basic substance. Cooking and eating oils and butter were almost universally used by people who prepared their food. Other common foods were not processed, merely ground, dried and cooked whole. Food was eaten where it was raised, or nearby, and it was preserved by drying, salting and by the natural refrigeration afforded by winter weather.

But when the food industry began a little over a hundred years ago with the widespread use of the tin can, it started a process of partitioning, fractionation, purification, mixing and preservation which has reached the highly specialized level it is today. Packaged, frozen, bottled, canned, wrapped food is nationally distributed and an enormous variety of products provided. We cannot begin to examine them, but we can explore the idea that all metallic micronutrients necessary for metabolism of a food fraction are or are not contained in that fraction.

In table X-1 are shown some foods and their fractions with the trace elements in them. Magnesium, the only bulk element which plays an active role in enzyme systems involved in the metabolism of foods, is also included. A careful examination of the table reveals:

TABLE X-1

Essential trace metals in partitioned foods, ppm

Food	Magnesium	Chromium	Manganese	Cobalt	Copper	Zinc	Molybdenum
Wheat	1502	1.75	49.0	0.75	4.08	31.5	0.79
Patent flour	299	0.23	6.0	0.36	1.50	8.9	0.32
Gluten	18	–	2.25	1.46	9.63	48.5	–
Rice	1477	0.16	2.8	0.16	4.10	6.5	–
Polished Rice	251	0.04	1.53	0.10	3.04	1.6	–
Corn	664	0.05	1.31	0.36	1.80	18.2	–
Meal	269	0.08	2.05	0.87	2.13	9.0	–
Starch	22	–	0.34	0.55	1.25	0.8	–
Oil	5	0.47	1.00	0.15	2.21	1.6	–
Sugar Cane	190	0.10	1.75	0.03	1.00	0.5	0.13
Raw Sugar	94	0.30	1.18	0.40	3.35	8.7	0.0
Molasses	250	1.21	4.24	0.25	6.83	8.3	0.19
White Sugar	2	0.02	0.13	<0.05	0.57	0.2	0.0

	sugar starch protein	sugar starch fat	protein fat?	red blood cells (B12)	oxygen	carbonic acid protein alcohol lactate	uric acid
Raw Milk	102	0.01	0.19	0.06	0.19	3.5	0.20
Homogenized milk	–	–	–	0.03	–	0.3	0.02
Skimmed milk	96	<0.01	0.0	0.36	0.29	3.0	0.02
Butter	6	0.17	0.96	0.35	3.92	1.7	0.19
Egg	113	0.16	0.53	0.10	4.10	20.8	0.49
Yolk	101	–	0.88	0.12	2.44	35.5	–
White	103	–	0.43	0.06	1.70	0.3	0.12
Beef	283	0.09	0.05	0.52	0.90	56.6	0.07
Fat	5	0.23	0.03	1.94	0.80	1.5	0.0
Pork	209	–	0.34	0.17	3.90	18.9	3.68
Lard	5	0.07	0.98	0.20	2.56	2.0	0.0
Essential for metabolism of:							

1. Grains. The refining of wheat to patent flour (72% of the wheat) produces a white product with 20 percent of the magnesium, 13 percent of the chromium, 12 percent of the manganese, 50 percent of the cobalt, 37 percent of the copper and 21-31 percent of the zinc in the whole grain. We repeat these data for emphasis.

Polishing rice results in a white product with 17 percent of the magnesium, 25 percent of the chromium, 73 percent of the manganese, 62 percent of the cobalt, 75 percent of the copper and 25 percent of the zinc in the whole grain.

Milling corn into meal removes 60 percent of the magnesium.

Although all essential metals but cobalt and molybdenum are probably involved at one stage or another of the degradation and build up of starches in grains, magnesium and chromium are especially required. Are there enough of these metals left in the refined product to supply the enzyme systems which take part in the metabolism of grains? We already know (Chapter VI) that there is not enough chromium and (Chapter V) that there are not enough of certain B-vitamins, for some of the latter have to be added. There is probably enough magnesium in flour, but not in gluten; the healthy body conserves it well.

2. Sugar. Refining of sugar produces a white product with 1 percent of the magnesium, 7 percent of the chromium and 12 percent of the cobalt in raw sugar. Magnesium and chromium are needed to metabolize sugar. Molasses, which is fed to cattle, is enriched in magnesium, chromium, cobalt, copper and zinc.

3. Oils and fats. Separation of oil from corn and refining it results in a yellow product with less than 1 percent of the magnesium, most of the chromium, manganese and copper and 25 percent of the zinc. Most of the vanadium is also in the oil.

Partitioning butter from milk produces a yellow product with 6 percent of the magnesium, most of the chromium and manganese, cobalt and copper and half the zinc. Skimmed or fat-free milk does not lose magnesium, copper or zinc, but has virtually no chromium or manganese, which are needed to metabolize sugar or protein. Thus a mixture of skimmed milk and corn oil is balanced in respect to trace metals, but neither one alone is adequate. Whole milk presumably is balanced, at least for calves.

Removing the fat from beef and pork and refining it produces a white product with 2 percent of the magnesium; a third of the chromium, 4 percent of the zinc, 20 percent of the copper and less than half the cobalt. Recombining these fractions results in a balanced diet in respect to these elements.

Egg yolk contains the same amount of magnesium, twice the manganese, twice the cobalt, 41 percent more copper and a hundred times the zinc in egg white. By recombining them, the trace metal content is presumably adequate for metabolism.

If these fractions contain inadequate amounts of micronutrients, there is good reason for people eating them to balance their diets with either whole foods or by going to other foods which contain large amounts of the missing ones. The second choice might still provide inadequate amounts unless there were surpluses in the other foods.

In Chapter V we have given the average amounts of the essential trace metals in various types of foods. We do not know if it is necessary to take the proper proportions of the trace elements when sufficient amounts are supplied. There is reason to believe, from experiments on sheep, that copper and molybdenum interact—a surplus of molybdenum will produce depigmentation of the hair of black sheep when copper intake is low. (Black and white striped wool can be grown by alternating the copper intake in black sheep grazing on a pasture high in molybdenum.) This interaction has not been shown in man.

The most logical way to insure adequate amounts of essential trace elements is to see that major sources of calories come from whole foods, unrefined foods or partly refined foods. It is impossible to exist today entirely on whole foods and enjoy a variety in one's meals. Therefore one must make compromises. In table X-2 is shown food consumption by class of food. We do not know that the average American diet provides levels of essential trace elements insufficient for optimal function but sufficient for survival, except in the case

TABLE X-2

Dietary intakes of foods by class in the United States, gram per day and available supplies.

Food	available g	ingested g	% of total calories g	g	adequacy of trace elements
Fish and Sea Food	26	22	1.4	0.86	Usually adequate
Meat and meat products	248	206	12.5	15.68	Usually adequate, except fats.
Milk, as liquid	850*	508	31.7	11.76	Whole adequate, skimmed deficient.
Eggs	55	47	2.9	3.13	Whole and yolks adequate, white deficient.

Cheese	– *	19	1.2	3.13	Usually adequate
Fats	56	49	3.4	17.64	Deficient in magnesium and zinc.
Sugar and preserves	113	69	4.3	10.98	All deficient in white sugars.
Potatoes	– †	103	6.2	2.74	Adequate with skins
Other vegetables	– †	202	12.5	3.13	Usually low
Fruit	652†	184	11.5	3.52	Usually low
Cereals	185	207	12.5	27.45	Refined low in all, whole adequate.
Total	2185	1611	100.0	100.0	

*Included in milk.
† All vegetables and fruits
Note: % of total calories in a 2550 calorie diet.

of insufficient chromium. But we can suspect, from experiments on animals, that this may be so. Because the essential elements are non-toxic except in large amounts, no harm can come from relative excesses, which other mammals thrive on.

Man is the only animal who cooks his food. Nutritionists have long suspected that much of the bulk and trace elements are cooked out of vegetables by boiling and thrown out with the water. In table X-3 are shown examples of vegetables raw, cooked and canned and of white and "whole wheat" bread. Losses of magnesium amounting to a half to three quarters of the original in the raw food were found in the cooking water of three vegetables. Two acid vegetables cooked in stainless steel pots gained chromium from the pot. Canning caused losses of manganese from spinach, but not from beets or asparagus. Cooking or canning produced losses of cobalt of a third to nine-tenths in five raw vegetables. We have no data on copper, but cooking or canning was responsible for serious losses of zinc in three of four vegetables. For all metals but manganese, whole wheat bread had much more than white bread; whole wheat bread is not usually made from whole wheat, but from an impure flour lacking bran and some of the germ.

In view of these losses, which can be prevented by steaming, baking and broiling, we cannot count on vegetables to provide us with surplus quantities of essential trace elements. Vegetables are low in everything but magnesium, and except for the starchy ones and legumes, provide few calories for energy. But we can examine the types of foods in table X-2 to see what practical suggestions can be made for insuring adequate supplies.

1. Meats and fish. Animal muscle usually is adequate in essential trace elements, and screens out non-essential ones. Recombined meats and fats such as we get in salami, other sausages and "luncheon meats" must be viewed with suspicion. Liver is high in trace elements, both essential and toxic.

2. Milk and cheese. Skimmed, low fat and fat-free milk is low in the trace elements which are removed with butter, although there is enough magnesium, copper and zinc.

Probably, whole milk is a better food than either butter or skimmed milk. In any case, skimmed milk should be covered by an oil or fat. Cheese is usually whole milk.

3. Eggs. White of egg is probably inadequate in respect to chromium, copper, and especially zinc. Whole eggs are a complete food, containing much fat and cholesterol in the yolks; chicken embryos grow and hatch on what the egg contains.

4. Fats. All vegetable oils contain little chromium, including highly refined ones. Beef fat and lard, saturated fats, have some chromium, perhaps because the beef are force-fed on a high carbohydrate—corn—diet before slaughtering. All fats and oils have plenty of copper, but are severely deficient in magnesium and partly deficient in zinc. Oils are believed to be a necessity for other reasons—vitamins A and D, for example, but it would be wise to cut the use of lard and saturated vegetable oils to a minimum.

5. Vegetables. Potatoes are probably adequate in essential trace elements if the skins are eaten. Most of the elements are in the skin and eyes, from which roots and shoots grow; the rest of the potato is largely carbohydrate to supply nutrients until roots can take over and extract them from the soil. Green vegetables usually contain adequate amounts of magnesium, manganese, copper and zinc, although many are deficient. Legumes generally are adequate. Examples of low manganese vegetables are carrots, onions, lima beans, string beans, green pepper, squashes, asparagus, cucumbers and tomatoes; of low zinc vegetables, string beans, kidney beans, onions, beets, beet greens, radishes, tomatoes, green pepper and asparagus; of low copper vegetables, green peas, yams, beets, green peppers, cucumbers, egg plant, asparagus, celery, parsley and rhubarb. All vegetables are low in cobalt.

6. Fruit and Nuts. Some fruits are low in copper—coconuts, for example. Apricots, papaya and bananas are low in zinc. Oranges, tangerines, pears, apples, apricots and canteloupes are low in manganese. We cannot depend on fruits to supply the trace elements, except for chromium which occurs in sizeable

TABLE X-3

Examples of losses or gains in essential trace elements in foods raw, cooked and canned, ppm

Food	Magnesium	Chromium	Manganese	Cobalt	Copper	Zinc
Tomato, raw	–	0.01	–	0.06	–	0.6
cooked	–	0.14*	–	0.04	–	0.2
canned	–	–	–	–	–	0.1
Spinach, raw	–	–	7.77	0.34	–	2.2
canned	–	–	1.42	0.10	–	1.3
Carrots, raw	185	–	–	0.10	–	–
cooked	62	–	–	–	–	–
canned	–	–	–	0.03	–	–
Celery, raw	66	–	–	–	–	–
cooked	31	–	–	–	–	–
Parsnips, raw	456	–	–	–	–	–

cooked	126	–	–	–	–	–	–
Beans, raw	–	–	–	–	–	–	31.5
canned	–	–	–	–	–	–	12.6
Beets, raw	–	–	0.41	0.06	–	–	0.5
canned	–	–	1.34	0.02	–	–	0.8
Asparagus, raw	–	–	0.32	–	–	–	-
canned	–	–	0.27	–	–	–	–
Beans, green	–	–	–	0.09	–	–	–
canned	–	–	–	0.01	–	–	–
Rhubarb, raw	–	0.02	–	–	–	–	–
cooked	–	0.05*	–	–	–	–	–
Flour, white	299	0.23	5.20	0.36	1.5	–	3.5
Bread, white	204	0.14	1.78	0.11	0.19	–	1.2
Bread, whole wheat	340	0.49	1.43	–	0.63	–	5.3

*Cooked in stainless steel

amounts in fruits and fruit juices. Whether or not there is extra chromium above the amount needed for the sugar in the fruit is not known.

All nuts are oily seeds, and as seeds they contain relatively large amounts of chromium, manganese (20 ppm), iron, cobalt, copper (15 ppm) and zinc (34 ppm). Thus, they represent an excellent source of essential trace elements, unsaturated fats and calories.

7. Cereals and grains. All grains are seeds, and as such they contain adequate amounts of the trace elements necessary for their growth—and for the health of mammals. As we have repeated *ad nauseam,* all highly refined flours are deficient. Whole rye flour is also deficient in chromium, enough so that rats fed it become deficient, developing diabetes and elevated cholesterol levels in blood. We should avoid white flour and its products, such as saltines, pie crust, cookies, biscuits and white bread, and use as unrefined a flour as we can get. For example, cake has 5 per cent of the manganese, 18 per cent of the copper, 11 per cent of the zinc and 10 per cent of the magnesium in the wheat from which it is made; crackers have 18 per cent of the manganese, 36 per cent of the copper, 22 per cent of the zinc and 18 per cent of the magnesium in the wheat. Flour with the germ still in it will not keep except in the cold, for its highly unsaturated oils oxidize at room temperature and become rancid.

8. Sugar, candies, preserves. White sugar is deficient in all of the trace elements necessary for its metabolism, without exception, and should be avoided. Raw sugar when imported is likely to be quite refined, with a low ash content, for there is some customs gimmick to protect the domestic sugar industry. Dark brown sugar has a high ash content, a rich taste, is high in chromium and other elements, and should be used exclusively. It does not flow freely, being sticky with molasses. The amount of ash, which contains the elements, is 3-5 percent for true raw sugar, 3 percent for dark brown, 0.3-0.5 percent for imported "raw" sugar, and 0.11 percent for refined white sugar. If one has a sweet tooth he can bathe it in artificial sweeteners, as long

as they are declared to be non-toxic; the cyclamates are now under suspicion and should be avoided by people with high blood pressure.

Probably the worst combination in respect to chromium is a lunch of salami on white bread, a sweet pie, and coffee with two or three spoons of white sugar. This type of daily meal should deplete chromium stores and raise blood cholesterol nicely.

There are many abnormal trace elements in foods, either as natural contaminants of the soil, or introduced as artificial contaminants from the products of civilization. We will concern ourselves only with those which have been proven toxic in more or less degree to mice and rats exposed for a lifetime; in order that we may avoid them, we need to know where they occur. We have already discussed this matter in Chapter VII.

The metals found toxic in some degree to rats and mice are lead, cadmium, germanium, selenium (in one form), tin, antimony, tellurium, and to mice in addition scandium, gallium, rhodium, indium. These last four are rather scarce, and there is no information on amounts in food, but it must be small. Nor is there evidence of industrial contamination. Germanium occurs in coal and "thousands of tons are discharged into the air daily with the smoke of coal and oil," according to Vinogradov, a pioneer in geochemistry. We have no evidence that air polluted with germanium will harm man; it has been little studied. Present efforts to clean the atmosphere, although pitifully small, show much promise because the citizenry is aroused and injury from smoke is suspected. Germanium is fairly evenly distributed in foods, although low in oils, the measured daily intake is 1.5 milligram, and it is readily excreted in the urine.

Avoiding large amounts of tin is fairly easy. It is not very toxic, but if one does not want it, one can avoid acid and oily foods in tin cans, especially when they are old. Today American cans are lacquered, and much less tin is absorbed into food than from unlacquered cans. The lacquer can be readily seen as a yellow coating; unlacquered cans are bright and silvery. Many foreign cans are uncoated. Tin is added to asparagus in glass so as to make it taste like canned asparagus! Coated cans have 3-12 ppm tin in foods, uncoated cans 4-129 ppm. There is tin in both iron and copper pipes, largely from solder, and there is tin in the air. The processing and packing of foods such as gelatin, smoked fish, macaroni, dried legumes, dried milk, milk in large cans and tea contaminate the food with tin. To illustrate this point, milk obtained directly from a cow's udder into polyethylene had 0.19 ppm tin; bulk milk in a large can from the same cows had 0.68 ppm, and dry skim milk had 0.96 ppm. Corn oil in cans had 4.1 ppm tin. We do not know for certain if tin in small quantities does any harm at all, but tin in large quantities probably should be avoided.

There is no reason to believe that antimony in present amounts causes any chronic disease, although it is toxic to rats and mice in terms of life span. We do not know how to avoid tellurium, unless we stay away from such foods as gelatin, butter, wheat, rye, noodles, almonds, hazel nuts, instant tea, molasses, spices, baking powder and yeast. It is doubtful that tellurium does any harm at these or larger exposures. Exposed workmen in metal refineries have a persistent garlic odor to their breath. There is little selenium in most foods; some grains grown in seleniferous areas have much. People living in those areas have poor hair and nails.

Some readers may ask how they can increase their intakes of the essential trace elements by supplements to the point of equalling the intakes of healthy laboratory animals. To do this is possible and safe, although governmental restrictions, and in some cases lethargy, have slowed or prevented the availability of trace substances in pills to physicians. Furthermore, few drug companies will package pills from the bulk chemicals available to all, if they cannot be patented.

In table X-4 are shown the amounts of essential trace elements in about the most potent vitamin and mineral supplement, Theragran-M (Squibb), the amounts which have been actually given to man over the long term, and the amounts on which laboratory animals thrive. The comparisons are interesting. It is clear that man has taken continually without

TABLE X-4

Essential trace elements in a vitamin supplement, tolerated by human subjects and fed to laboratory animals, per day.

Element	Theragran-M mg	Theragran-M % of need	Given to man mg	Given to Animals* mg	Form of supplement for man
Chromium	0	0	10	49	Chromic acetate or oxalate
Manganese	1	25	10	230	Manganese acetate or citrate
Iron	12	100	1000	400	Ferrous sulfate and others
Cobalt, as B_{12}	5 (μg)	500	10	0.8	Cobaltous acetate or chloride
Copper	2	57	?	40	Copper acetate or citrate
Zinc	1.5	12	150	200	Zinc acetate
Molybdenum	0	0	10	4.9	
Fluorine	0	0	5	50	1 ppm sodium fluoride safe
Iodine	0.15	150	100+	4.4	Potassium iodide

*Corrected for body weight and food consumption.

harm, 9 times his requirement or usual intake of chromium, three times that of manganese, 100 that of iron, a thousand or more that of cobalt, 12 times that of zinc and, from natural sources, 5 times that of fluoride.

Amt's recommended

A logical, reasonable and ideal therapeutic capsule, based on the findings reported in this book, would contain 10 mg manganese, 5 mg chromium, 30 mg zinc and 0.5 mg fluoride. One could add iron, as is popular,* but cobalt and copper can be neglected. Iodine in addition would do no harm. Information on molybdenum is lacking. This combination would insure intakes of essential trace elements adequate even for the aged.

At this point, based on experiments in rats and mice, we can offer a theory of longevity of life span of man. Just as when a person is exposed continuously to harmful doses of radiation, which shorten his life non-specifically, so could the slow accumulation of abnormal and toxic trace elements shorten his life. We have seen how deficiency of chromium can lead to atherosclerosis, which definitely cuts life short. We have seen how slow accumulation of cadmium can lead to hypertension, which definitely cuts life short. These two conditions existing together can be more lethal than either alone. Today we can say that accumulation of lead eventually leads to some sort of metabolic breakdown which sooner or later results in ill-health and perhaps death.

Barring the ill-luck of cancer, accidents, infections, and other less common diseases, we can predict that persons having little cadmium in their tissues and adequate amounts of chromium in their tissues, and having a low body burden of lead, may well survive in reasonable health to their allotted life spans of 90 to 110 years. Man always has lived to this age

*It is likely that self-medication with iron and iron-containing tonics by middle-aged persons causes a sizeable number of deaths from intestinal and stomach cancer. Growing children and menstruating women on diets poor in iron may need it to prevent anemia. The commonest cause of iron-deficiency anemia in older persons is chronic bleeding from the gastro-intestinal tract, which may go unnoticed for many months. Iron will correct this anemia, but if the bleeding comes from a cancer, iron can cover up the symptoms until the cancer has grown to an inoperable size. Anemic older people with "tired blood" should consult a physician to find its cause and not treat themselves into a false security.

according to recorded history, although most of his fellows have died younger. With the application of suitable measures to control the trace elements which are so basic to optimal function, the lives of the majority may be allowed to reach maximal biologic age. So do the animals tell us.

CHAPTER XI

Pollution by Industrial Metals

Man has been unwittingly exposing himself to increased amounts of trace metals in food, water and air since he first learned to use metals—in fact, since he first used fire. Only for the past hundred years, however, has he been polluting his environment with metals in large amounts, from increasing industrialization, smelting, refining and the burning of coal and petroleum for energy.

Pollution by metals and other elements must be considered in terms of the whole environment, air, water, food and in some cases, contact. What goes up in smoke must come down in soot, rain or snow, entering the soil and plants in the food chain, entering the waters to be taken up by plants and fish, entering the human body by breathing contaminated air, drinking contaminated water, eating contaminated food. Certain metals are easily absorbed by the human lung or intestines; others are poorly absorbed, accumulating in lungs with age but not accumulating when taken in food or water. Others accumulate in the human body with age and continuous exposure, and some of them can cause disease in older persons thus exposed. The important questions are: Which elements are absorbed from lungs? Which accumulate in human tissues? What are the signs of innate toxicity, if any? Therefore, we must examine metallic pollution sensibly and in the light of knowledge as to the behavior of each pollutant. The magnitude of possible exposures and the innate toxicities are shown here in Tables XI-1 and 2.

AIR POLLUTION BY METALS

There are at least 27 trace elements found, or probably found in air. Five are natural pollutants from soil and dust; exposure to them is age old. Four of these, *titanium, aluminum, barium* and *strontium,* accumulate to some extent in the human lung, but none is toxic, as far as is known. The fifth, *iron,* is both a natural and an industrial pollutant; it is absorbed by the lung, is essential for life and seldom accumulates in the body. These five can be neglected at present levels of exposure.

Six other metals found in air are essential: *chromium, manganese, cobalt, copper, zinc, molybdenum.* Chromium is an industrial pollutant from coal, accumulating in insoluble particles in the lung; natural chromium occurs in food. The others do not accumulate, and merely add small increments to the needed amounts absorbed in food. They can be neglected at present levels of exposure.

The remaining 16 are industrial contaminants. The burning of coal provides air polluted with nickel, silver, titanium, antimony, beryllium, vanadium, fluorine (essential as fluoride for health), lead, boron, mercury, germanium, gallium, tin, tungsten, arsenic, selenium, zirconium, yttrium, bismuth and probably more, for coal has concentrated in it all the elements present in primitive trees. The burning of petroleum and its products (except gasoline and distillates) provides air polluted with titanium, nickel, arsenic, barium, vanadium, niobium, chromium, zirconium, tin, yttrium, gallium, scandium, uranium. Cadmium, tin, antimony and bismuth are also found in air from various industrial sources. Airborne lead is largely a result of anti-knock additives to gasoline, in the form of tetra-ethyl or

TABLE XI-1

Annual U.S. Consumption of metals (1968) and their effects on mice and rats exposed for life in drinking water at 5 ppm or more

Metal	Estimated Consumption, Thousands of Tons	Dose, ppm	Innate Toxicity Mice Growth	Innate Toxicity Mice Life Span	Innate Toxicity Rats Growth	Innate Toxicity Rats Life Span	Remarks
Aluminum	3,644	10	0	—	0	—	Inert
Manganese	2,228	10	0	0	0	0	Essential
Zinc	1,728	50	0	0	0	0	Essential
Barium	1,590	5	0	—	0	—	Sl. toxic
Copper	1,576	5	0	0	0	0	Essential
Lead	1,329	25	0	+	0	+	Toxic
Chromium	1,316	12	0	0	0	0	Essential
Nickel	159	5	0	0	0	0	?Essential
Titanium	89	5	0	0	—	—	Inert
Arsenic	25	5	0	+	0	0	Toxic
Beryllium	9	5	0	—	+	—	Toxic
Cadmium	7	5-25	0	+	+	+	Toxic
Vanadium	5	5	0	0	0	0	Essential
Silver	5	—	?	?	?	?	Non-toxic
Mercury	3	5	0	—	0	—	Toxic
Selenium	0.8	3	0	0	0	0	Essential (Cancer)

Note: All metals are toxic in large enough doses.

tetra-methyl lead. Mercury in paints and in seeds can volatilize into air.

Six of these elements can be hazards to human health. Of these, cadmium is a present and real hazard, as we have pointed out. Lead offers a potential and imminent hazard. Nickel carbonyl from residual oil and coal has caused cancer of the lung in animals and exposed workers, and therefore presents a potential, if not a real hazard to the public health. These three require immediate control.

Beryllium has appeared in the air of nine urban and four non-urban areas of the United States. It accumulates in lung, can cause cancer of the lung and serious disease; while probably a minor hazard at present, it could become a real one if not carefully controlled.

Antimony in low doses shortens the life span and longevity of rats; it has appeared in the air of seven areas, and if not controlled, could become a hazard.

Mercury in air is probably of little importance to health at present, unlike mercury in water.

Cadmium is a constant contaminant of zinc, and where there is zinc (an essential metal) in air, there is cadmium, in ratios from 11:1 to about 48:1, depending on the area. Air, however, probably provides only a small increment of the total cadmium daily absorbed, the largest part coming from water and food.

Lead from motor vehicle exhausts enters the environment in amounts of two pounds per capita per year. We have found enough lead in vegetation growing beside a secondary highway (up to 200 parts per million, wet weight) to abort a cow subsisting on this vegetation; the concentration has trebled in six years. Fifteen of 20 samples of melted snow from the same place had more lead than the allowable limit for potable water, and seven samples had more than five times that limit. Horses have died of lead poisoning from contaminated hay, in England and in California. Evidence of a biochemical abnormality in persons exposed to urban air concentrations of lead is beginning to appear. There is little doubt that at the present rate of pollu-

TABLE XI-2

Annual U.S. Consumption (1967) of metals produced in small quantities and their effects on mice and rats exposed for life in drinking water at 5 ppm or more

Element	Consumption Tons	Innate Toxicity			Remarks
		Dose, ppm	Growth	Life Span	
Magnesium	89,300	–	0	0	Essential
Tin	80,646	5	0	+	Sl. toxic
Zirconium	60,909	5	0	0	Inert
Lithium	25,000	–	–	–	Sedative
Molybdenum	22,502	50	0(R)	0	Essential
Antimony	16,682	5	0	+	Toxic
Cobalt	6,352	1	0	0	Essential
Tungsten	6,210	5	0	+(R)	Toxic

Uranium	4,900	—	—	—	Toxic
Germanium	4,091	5	+(M)	+	Sl. toxic
Niobium	1,451	5	0(M)	0	Inert
Bismuth	1,143	—	—	—	Toxic
Boron	451	10	0	0(R)	Essential, plants
Thorium	120	—	—	—	Toxic
Strontium	107	5	0	0	Essential
Tellurium	78	2	+	+	Toxic
Platinum	19.7	—	—	—	Inert?
Palladium	19.3	5	+(M)	0	Carcinogenic
Indium	1.9	5	0(M)	0	Inert
Rhodium	1.7	5	+(M)	0	Carcinogenic

(R) = Rats
(M) = Mice

tion, diseases due to lead toxicity will emerge within a few years. There is urgent need at present to reduce and eventually eliminate lead additives to gasoline, before environmental lead builds up much further.

Nickel carbonyl is formed when hot carbon monoxide passes over finely divided nickel. Coal, petroleum, and residual oils contain nickel, which is extruded in chimneys and exhaust pipes in an atmosphere of hot carbon monoxide. Methods are available to reconvert the resultant nickel carbonyl, which is retained in lung and is carcinogenic in the lungs, to nickel or nickel oxide, which is non-toxic and is not retained in human lungs. Nickel salts in low doses may be essential for birds and possibly mammals. We have calculated that the annual global emission of nickel from fossil fuels into the air amounts to 70,000 tons or 14.5% of annual world production, worth some 110 million dollars.

Vanadium, a component of Venezuelan and Iranian low sulfur petroleum (there is less or little in other known oils with higher sulfur content), accumulates in human lungs with age. Vanadium in air occurs in a belt along the Atlantic Seaboard and in Puerto Rico, where these oils are directly delivered, and in the North there is more in winter air than in summer. In low doses it had no effect on longevity of mice and rats, and probably presents no real and present hazard to health. *Niobium* also occurs in petroleum; it also showed little or no innate toxicity.

Arsenic occurs in coal and oil. In spite of its reputation as a poison, it has a low order of toxicity to mammals, and in small doses for life has no detectable biological effects. It is well tolerated, there are relatively large amounts in seafood, and it is doubtful that arsenic is innately toxic at present levels of exposure. (In fact, it promoted longevity of rats.) In man, however, it causes skin lesions which can result in cancer.

Selenium is a constant contaminant of sulfur, and therefore occurs in the air probably as the dioxide, along with sulfur dioxide. It is carcinogenic in rats; whether the low concentrations in polluted air cause prolonged ill effects is not known.

Selenium in very small doses is essential for mammals. Human intakes from air have not been measured.

Germanium has little innate toxicity, and *boron* or borate probably has none, although borates are toxic. *Yttrium* is carcinogenic in mice exposed to it for life; it has not been measured in air. Also carcinogenic are *rhodium* and *palladium.*

Therefore, there are five trace metals in urban (and non-urban) air which are of concern in respect to human health, and 22 which are of little concern at present levels of exposure. A number of others have not been measured.

WATER POLLUTION BY METALS

The U.S. Geological Survey has analyzed municipal water supplies of the 100 largest cities of the United States. Twenty-three trace elements were looked for, 16 were found in sizeable quantities. Of these 16, five are essential, 10 are biologically inert, and only one, *lead*, is toxic on lifetime exposures.

The maximum concentration of lead found was 62 parts per billion, which exceeds Federal Criteria of 50 ppb; the median concentration was 3.7 ppb. Lead in potable water has offered a hazard to health for over 2000 years. Only about 10% of the lead ingested in water, however, is absorbed by the body; nearly half the amount absorbed by urban dwellers comes from air. Lead from motor vehicle exhausts can contaminate water in reservoirs to a small extent. Lead in water mainly comes from lead pipes containing soft, acid water.

Cadmium was rarely detected in water because the method used was fairly insensitive. By using a sensitive method, we found that eight of 23 tap waters had cadmium exceeding allowable limits of 10 parts per billion, and demonstrated a pick-up of cadmium from water mains and pipes in houses compared to water in reservoirs and streams. This solution of cadmium from pipes occurs only in areas with soft, acid waters, not in areas having hard waters. Because there is a direct relationship of cardiovascular death rates in the United States,

Great Britain, Sweden, Canada and Japan with the degree of softness or acidity of water, we strongly suspect cadmium, which can cause high blood pressure, as the offending agent.

Some rivers and streams in the United States are contaminated with trace metals. The U.S. Geological Survey analyzed over 720 samples from 50 states and Puerto Rico. There was more than the allowable limit of *cadmium* in 4%, of *arsenic* in 2%, and of *lead* in a few waters. One South Carolina river had 1.1 ppm arsenic (the maximum allowable for drinking is 0.05 ppm). The highest value of cadmium was 130 ppb, (10 ppm allowable). *Chromium, cobalt, mercury*, and *zinc* were below allowable limits. These results on arsenic and cadmium point out how local, but not general, contamination can occur.

Permissible criteria for limits of trace elements in water are mainly satisfactory, although from the standpoint of innate toxicity alone they could be raised in the cases of *barium, boron, chromium, iron, manganese*, and *zinc* and lowered for *cadmium*. No changes on this basis are recommended for *arsenic, copper, fluoride, lead, selenium*, or *silver*. Whereas municipal water supplies for large cities are generally satisfactory, those for small towns sometimes contain elements with possible toxicity in amounts exceeding drinking water standards.

Undoubtedly there are other trace elements dissolved in water which are not removed by treatment, but aside from lead and cadmium, none appear to be potentially hazardous to human health except under unusual circumstances and conditions. Analyses are needed, however, in the cases of polluted rivers containing factory effluents which supply municipal waters or which contain toxic elements when these elements enter the food chain.

An example is *mercury*. Total maximal potential exposure to mercurial compounds used in agriculture, paint and dental fillings amounts to 1.85 grams per capita per year. If pharmaceuticals are included, the total is 2.6 grams. Undoubtedly only a small amount of this total represents real exposure. Unlike lead, inorganic mercury does not accumulate in the body but organic mercury is deposited in fat and nerve tissue, where it is less mobile.

There probably has been mercury in fish ever since there were fish, for there is a background level in water. Mercury has been in air since the dry land appeared, for it is present in soils (up to 0.8 parts per million) and it is volatile. Mercury is found in all living things, but the increasing industrial use, from 2.7 million pounds in 1947 to 5.5 million pounds in 1966, has caused contamination of waters. As a result, fish have accumulated mercury. The amount of mercury used in agriculture and paper has decreased considerably, however, in ten years, while that in paint has tripled and that in drugs has doubled. Mercury enters the air from coal burning. Mercury as a source of water pollution should be controlled, in order to avoid possible toxic effects in human beings who eat much contaminated fish. The amount of mercury in water itself offers no direct hazard to human health when the water is drunk.

METALLIC POLLUTION OF FOODS

The relative amounts of essential trace elements absorbed from water are small compared to those absorbed from food. In respect to the toxic trace elements, nearly half of the total lead absorbed by urban dwellers comes from the air, and probably a sizeable increment of total cadmium comes from water. The potential hazards from lead, beryllium, nickel carbonyl and antimony today are to be found in polluted air; the hazard of cadmium is in both air and water. In respect to food, the situation is reversed.

Extensive analyses of foods and waters by our laboratory have revealed that, in general, most foods raised locally or bought in supermarkets or chain stores are relatively free of toxic metal contaminants. The food industry has kept its products quite clean in respect to contaminating trace elements, with individual exceptions. Canning introduces tin into foods, but it has a low order of toxicity.

The hazard in foods occurs not from contamination but

from refinement. Major sources of caloric energy are largely refined, and in refining, much or most of the trace metals essential for health are removed. Unfortunately, they are not restored to the food. Thus, at present, we have a situation in which domestic animals and pets receive more than adequate amounts of elemental micronutrients, whereas man can get only marginal amounts.

To recapitulate, the milling of wheat into refined white flour removes 40% of the *chromium,* 86% of the *manganese,* 76% of the *iron,* 89% of the *cobalt,* 68% of the *copper,* 78% of the *zinc* and 48% of the *molybdenum,* all trace elements essential for life or health. Only iron, and that in a form poorly absorbed, is later added to flour. The residue, or millfeeds, rich in trace elements, is fed to our domestic animals. And by the same process, most of eight vitamins are removed from wheat; three are added to make the flour enriched; millfeeds are rich in vitamins. Similar depletion of vitamins and essential trace elements occurs when rice is polished and corn meal is refined. Likewise, most of the bulk elements are removed from wheat: 60% of the *calcium,* 71% of the *phosphorus,* 85% of the *magnesium,* 77% of the *potassium,* 78% of the *sodium,* which appear in the millfeeds.

Refining of raw cane sugar into white sugar removes most (93%) of the ash, and with it go the trace elements necessary for metabolism of the sugar: 93% of the *chromium,* 89% of the *manganese,* 98% of the *cobalt,* 83% of the *copper,* 98% of the *zinc* and 98% of the *magnesium.* These essential elements are in the residue molasses, which is fed to cattle.

Refined fats contain little magnesium, cobalt or zinc but have adequate amounts of copper and manganese. They are generally low in chromium. Therefore, most of the energy in the average American diet, which comes from white flour, white sugar and fat, is not supplied with the trace substances needed to utilize that energy efficiently and properly.

Requirements of domestic and laboratory animals for essential trace elements are much higher than are the amounts found in American diets. Whereas specific deficiency diseases are not generally recognized, there is evidence accumulating that dietary deficiency of zinc may be fairly common in older persons, in pregnant women and in patients with liver disease and chronic infections. It is possible, but not proven, that human beings subsist on marginal intakes of manganese; animals given the same concentrations as are found in human diets show signs of deficiency.

Some abnormal trace elements cause cancer in animals when they are fed for a lifetime. Of 26 studied, we found only four: yttrium, rhodium, palladium (no problem of pollution from these metals, so far as is known), and selenium. There appears to be no contamination of food or waters by these elements, and they occur in very small quantities.

Strict adherence to and broad interpretation of the Delany Clause to the Food Additives Act of 1958 would require that beryllium and selenium be prevented from entering air and water, that environmental control of yttrium, rhodium and palladium be strict, and that no nickel carbonyl be emitted from exhausts and in smokes from nickel refineries.

INTERACTIONS

Life has evolved in the presence of a balance of trace elements found in sea water. Toxic elements are there in small concentrations; necessary ones are readily available. Plants on land obtain their necessary trace elements from soil, and within certain limits, reflect the soils on which they are grown. Animals get their trace elements from plants. Deficiency of an essential trace element causes disease in both plants and animals; high concentrations are toxic and lead to non-survival. Thus, when the balance is disturbed, from natural or man-made causes, the result can be serious.

Abnormal trace elements in larger than optimal amounts may enter living systems and interfere with the actions of essential trace elements. Similarity of atomic structures can lead to the interaction of one element with another in a system and cause biochemical alterations which lead to disease. This reaction occurs when the balance of nature is upset, as it is now that man has added large concentrations of metals and other elements to his environment by mining them and literally spreading them over the face of the earth in air and water.

Environmental metals may cause deficiencies of essential metals by interfering with chemical reactions basic to life or health. They have been little studied from this viewpoint, and much more research is needed. The importance of this problem is illustrated by the fact that rats and mice spending their lives in an environment controlled as to metallic contamination live 20-25% longer than do animals in the contaminated environment of the usual animal quarters. Furthermore, congenital abnormalities have been produced by exposing breeding mice and rats to cadmium, lead or selenium, for several generations, and the sex ratios of offspring have been altered by arsenic and molybdenum. There is some correlation of certain qualities of municipal water supplies and deaths from congenital abnormalities in the United States, and the qualities involve trace elements. This phase of the whole problem is only beginning to be studied.

I must emphasize that environmental pollution by toxic metals is a much more serious and much more insidious problem than is pollution by organic substances such as pesticides, weed killers, sulfur dioxide, oxides of nitrogen, carbon monoxide and other gross contaminants of air and water. Most organic substances are degradable by natural processes; no metal is degradable. Elements in elemental form or as salts remain in the environment until they are leached by rains into rivers and into the sea. Some of them are slow to move; lead and arsenic, for example. Those metallic and elemental pollutants we have with us now are here to stay for a long time. Therefore, every effort must be made to slow the environmental build-up of those elements which are toxic and can cause degenerative diseases, neglecting those which are essential for living things and those which are biologically inert.

In order to maintain the present environmental balance of trace elements, we must control further pollution by those elements which give us concern. In the order of importance, cadmium in air can be minimized or virtually abolished by abatement of zinc, from which it comes. Lead in air can be virtually abolished by eliminating alkyl lead additives to gasoline. Nickel carbonyl must be specifically treated to decompose it in smelter and refinery stacks and in emissions from chimneys and exhausts, and no nickel additives should be allowed in gasoline. Beryllium and antimony in air can be controlled by reducing particulate emissions from coal smokes. Mercury in water can be controlled by regulation of factory effluents and by finding less toxic fungicidal compounds for grains, paper and paint. If measures for abatement were directed at these six metals, particulate matter in air and pollution of water would inevitably diminish.

This discussion does not exclude the possibility that other trace elements insidiously toxic to man may be discovered in the environment on further intensive research.

CHAPTER XII

Serious Local Pollution
by Metals with
Effects on Health

As always happens when human activities are uncontrolled or unpoliced, or when persons pursuing these activities lack knowledge and foresight, something goes wrong and innocent people suffer. Industry is not alone in avoiding possible harmful effects of its activities and in the past has blatantly discharged metallic waste products into the air and water in large amounts, turning its corporate back on far-reaching consequences. Now industry is alerted fully to the probability of its having to pay the piper for its excesses by public pressure and governmental action. The honeymoon with waste disposal is over; it has lasted two thousand years.

The news media have now built up a national paranoia on unseen poisons in the environment which could ruin our health: pesticides, food additives, radiation, trace elements and the like. Some scientists have been affected by this mental disease, and add their strident cries of danger to that of the press; in many cases they start it off. There is an abysmal lack of reason, logic and common sense shown by some scientists which is reflected in the attitudes of the public. (I just received a letter saying that the writer had read some remarks of mine in *The New York*

Times: "Why did I not say something sensational?") The voices of those pleading for reason based on fact go unheard and their fate seems to be *Vox clamantis in deserto*.

Three trace metals which are wasted into the environment are today popular subjects of scares. These are lead, cadmium and mercury. Two of them have caused deaths in *strictly local* situations, both in Japan. Let us see what the facts are, remembering that any element, or any substance, in sufficient excess can be toxic. This fact is highly important. For example, cyclamates were barred from food and drink because large excesses caused cancer. Two Japanese pathologists who came to see me a year ago told me with a smile, "Yes, we injected cyclamates and produced cancer. We also injected sucrose (sugar) and it caused cancer." No one publicized *that* experiment! Water, salt, food in excess is toxic and can kill.

Cadmium toxicity. A large zinc mine and smelter for many years dumped its effluents into the air and into the water of the Jintsu River, Toyama Prefecture, Japan. Down river in a local area the farmers used the river water to irrigate their crops of rice, wheat and vegetables, and lived on their own produce. Fifteen to twenty years or so after the dumping began, some 200 old men and women started complaining of aches, pains, deformities of the spine and easily broken bones. The disease was extremely painful; it was named "itai-itai disease" (ouch! ouch!), and was eventually fatal in half the cases. Men were relatively less affected and children never.

What had happened was that the river water was heavily contaminated with soluble cadmium and lead salts from the mine. These metals entered the food of the local people and slowly accumulated in their bodies. Cadmium has little or no affinity for bone, but lead has. Japanese farmers live on low calcium diets, about half the U.S. average, drinking almost no milk and eating few dairy products. There was a great deal of cadmium and lead in the bones and organs of the few women dying and subject to analysis (Table XII-1).

The disease has not been seen in Europeans exposed to

TABLE XII-1

Cadmium and lead in human tissues from a patient with
Itai-Itai Disease, ppm ash.

Tissue	Patient		Normal American, av.	
	Cadmium	Lead	Cadmium	Lead
Rib	11,472	408	0	43
Cartilage	4,755	156	0	0.5
Vertebra	6,967	260	0	54
Lung	2,554	847	35	47
Kidney	4,903	999	2,900	98
Liver	7,051	229	180	130
Brain	293	342	0	5
Stomach	3,762	661	< 4	12
Small Intestine	4,084	2,215	4	20
Soil, paddy (5 samples)	6	348	0.5	5
Roots, rice (5 samples)	1,250	810	—	—
Rice (17 samples)	125	22	3.2	2

cadmium in cadmium-nickel battery factories, only in this one area of Japan and one other. Along the Hudson River a similar situation could develop *if* people irrigated their gardens in a local area with water from a certain cove, *if* there were no supermarkets, *if* there was extensive lead pollution of the water also, and *if* people drank the water exclusively and lived on the food they raised.

A factory making cadmium-nickel batteries discharged its wastes into Foundry Cove on the Hudson River, apparently for many years. Observers found that the cove bottom had a yellow tinge, and sent us some mud. The analyses are shown in Table XII-2. When asked what we recommended, we said that we would mine it for cadmium and nickel; it had 22.6% nickel. A ton of dry mud would be worth $339 worth of nickel at 75 cents a pound and $766.20 of cadmium at $2.55 a pound, or about $1,100 a ton. In other words, these pollutants could be recycled at a profit.

Fish in the cove took up cadmium to the point where it

could be toxic to human beings eating minnows regularly. A 20 lb carp also took up cadmium, although it was caught several miles down stream. Pollution went a long way.

Naturally we want to know whether or not fish living in less contaminated rivers also take up cadmium. Three rivers out of four in Alabama have more cadmium than allowable in drinking water by the Public Health Service, 10 ppb. In Table XII-3 are shown the results. A little more cadmium was found in fish living in 90 ppb water than 65 ppb, 12 ppb and 6 ppb water. But the levels were not dangerous or hazardous to health. I would not drink the water, but I would eat the fish.

These two examples indicate that we must not be scared by heavy pollution in one area by applying it to slight contamination in other areas—or to all parts of the country.

Methyl Mercury. Again in Japan, from 1953 to 1960, a factory making plastics dumped a catalyst, methyl mercury, into Minimata Bay. The methyl mercury was taken up by fish and shellfish, up to 50 ppm in fish and 85 ppm in shellfish. The local population subsisted largely on local fish, eating an estimated amount of half a pound a day, plus shellfish. A hundred and thirty-four became ill after several months, and 48 died. Children were born with small brains; they were imbeciles. Sufferers had many nervous symptoms of bizarre natures. Many fish and some farm animals fed fish died. When the cause was suspected, the Japanese did a great deal of work on it, reproducing the disease in rats and cats by feeding them local fish. The factory was made to stop its pollution and no more cases appeared. The same disease broke out in 49 people four years later in Niigata, from the same cause and with the same cure. Six died. The disease is permanently disabling.

It was interesting that the bottom mud just outside Minimata Bay had no more mercury than background levels. Thus, this severe pollution with methyl mercury and probably inorganic mercury was confined to the bay. There were no cases in other parts of Japan, or anywhere else in the world, from eating fish.

TABLE XII-2

Cadmium in fish from a polluted cove and downstream in the Hudson River.

	Cadmium, ppm
Foundry Cove	
Mud	162,000(16.2%)
Water	2
Cattail, unwashed	667
washed	226
Minnows	
2 silver dace, whole	9.0
3 silver dace, whole	7.6
6 small bass, whole	14.1
10 small bass, whole	11.2
silver dace, flesh	2.5
heads	2.1
innards	11.8
bass, flesh	4.0
heads	5.1
innards	53.5
Hudson River, Montrose, N.Y.	
downstream from Cove 12 miles	
20 lb. carp	
Flesh, left back	0.04
right back	0.67
Innards	
Liver	3.36
Liver, degenerate	12.32
Kidney	20.44
Heart	0.30
Intestine	0.83
Roe	1.20
Bone	0.40
Fat	0.13
Gill	0.39
Scales	0.19
Hudson River, near Montrose, N.Y.	
Pumpkin seed Sunfish	0.24
White Perch	0.25

TABLE XII-2 (Continued)

	Cadmium, ppm
Hudson River (Continued)	
Herring	0.16
Striped Bass	0.25
Snapping Turtle Eggs	0.05
Miramichi River, N.B.	
Atlantic Salmon, liver	0.26
Atlantic, North	
Halibut, steak	0.05
Haddock, fillet	0.09
Fish flour, ground fish	1.56
Dried mussels, Calif.	4.93

Shell fish can take up a number of metals from sea water, oysters especially (Table XII-4). Zinc, cadmium, copper and iron are high in some oysters, although the ratio of zinc to cadmium is 460-530 to 1. We do not know if the zinc covers the cadmium, making it harmless. Some oysters and clams take in much lead, probably from polluted waters, enough to cause lead poisoning if eaten in large quantities.

The disease, methyl mercury poisoning, has occurred, however, in people eating grain treated with methyl mercury fungicides, as might be expected. There are many cases of poisoning of all kinds in people who can't or don't read warning labels. In Iraq in 1956, more than 100 people were poisoned by an ethyl mercury compound in flour from treated wheat; 14 died. In 1960, also in Iraq, several hundred more were poisoned, many died. In West Pakistan in 1961, more than a hundred were poisoned by flour from wheat seed treated with ethyl mercury chloride to prevent molding; at least 4 died. In Guatemala from 1963 to 1965, about 45 people were poisoned by Panogen, a methyl mercury fungicide used to treat wheat; 44% died. In Alamogordo, New Mexico, in 1969, three children of nine became ill with methyl mercury poisoning and a baby was born

TABLE XII-3

Cadmium in fish from contaminated Alabama Rivers, ppm wet weight

	Shell Bayou		Black Warrior		Cahaba		Mobile		Tennessee	
	No.	Conc.	No.	Conc.	No.	Conc.	No.	Conc.	No.	Conc.
Water, Cd ppb	—			6		12		65		90
Fish										
Bass and bluegill	(4)	0.02	8	0.03	2	0.01	5	0.02	8	0.08
Redear	4	—	—	—	—	—	4	0.02	1	<0.01
Drum	—	—	1	0.03	—	—	—	—	—	—
Catfish	—	—	—	—	2	0.02	4	0.02	—	—
Sunfish	—	—	—	—	5	0.04	—	—	—	—
Totals and Means	4	0.02	9	0.03	9	0.04	13	0.02	9	0.07

Note: The average value in the Tennessee River fish was significantly larger than that in the Mobile River fish, but not significantly larger than those of the other two river groups.

TABLE XII-4

Trace metals in shellfish, wet weight, ppm

| | Oysters | | | | Hard Clams | | Soft Clams | |
| | Atlantic | | Gulf | | Atlantic | | | |
Metal	Ave.	Max	Ave.	Max	Ave.	Max	Ave.	Max.
Essential								
Chromium	0.40	3.4	0.33	3.3	0.31	5.8	0.52	5.0
Manganese	4.30	15.0	—	—	5.8	29.7	6.70	29.9
Iron	67	238	57	65	30	1,710	405	1,710
Cobalt	0.10	0.2	—	—	0.20	0.28	0.10	0.2
Copper	91.5	517	25.8	138	2.6	16.5	5.0	90
Zinc	1,428	4,120	516	3,800	20.6	40.2	17	28
Toxic								
Cadmium	3.10	7.8	0.88	2.6	0.19	0.73	0.27	0.9
Lead	0.47	2.3	1.11	17.3	0.52	7.5	0.70	10.2
Nickel*	0.19	1.8	—	—	0.24	2.4	0.27	2.3

*Non-toxic orally.
Data from Food and Drug Administration, Shellfish Sanitation Branch.

probably poisoned. A hog fed seed treated with Panogen was killed and eaten by the family. The children are probably permanently crippled.

These episodes, all the result of methyl mercury poisoning, have provided the basis of a national scare about mercury poisoning almost to the point of paranoia. Those who contribute to the scare forget certain basic facts which argue against mans' contaminating his environment with mercury, and especially with methyl mercury, other than in the local areas.

1. Mercury is in every living thing and in every natural bit of matter, be it land, or sea or air.

2. Mercury has been in fish ever since there were fish. (Since the third day of Creation; according to Genesis.)

3. Mercury slowly accumulates in vertebrates, and probably in invertebrates, with time and age.

4. Fish eaters, be they fish, flesh or fowl, have more mercury in their organs and tissues than non-fish eaters. The older the fish, the larger, and the more mercury it has. This explains why swordfish and tuna have more mercury than cod and halibut, and why carnivorous pike have more than whitefish or suckers.

5. The ocean contains 1.42 quintillion tons of water. The few thousand tons of mercury added to it yearly from natural run-offs and man's activities could not be detected in a hundred years. Besides, mercury in river and ocean water, like most metals, is scavenged by sediment and falls to the ocean bottom to become ore. The seas have not been contaminated by man.

6. Mercury in ocean fish is background mercury, regardless of the amount—unless the fish have fed in polluted waters of estuaries or bays. No one has shown high levels of methyl mercury in ocean fish.

7. Fish preserved since 1927 have shown mercury levels up to 1.03 ppm total mercury, more than twice the allowed level of 0.5 ppm. Swedish bird feathers more than 100 years old have had high mercury levels. Little mercury was used then, and seeds were not treated.

8. Treated seeds do not appear to contribute significantly to soil mercury levels, which vary widely over the United States.

9. No one has become ill from eating fish containing mercury except where methyl mercury was dumped.

10. All forms of mercury, except alkyl (methyl and ethyl) mercury, are only moderately toxic. Man has been ingesting mercury in fish for as long as he has fished.

The other side of the argument is based on these steps, which have several potholes:

Step A. When mercury is discharged into a river or a lake, or presumably into the ocean, certain bacteria not requiring oxygen will methylate it. Thus it is converted from a non-toxic to a toxic form, which enters the food chain of fish.

This conversion was demonstrated in mud from an aquarium at 75°F in the laboratory, in the presence of high concentrations of mercury in water. Lake mud was less active. There is no proof that this conversion occurs widely in natural cold water, nor at all in sea water. If it did, bottom feeding fish would be expected to have more mercury than other fish. They have less.

Step B. Most of the mercury in fish is methyl mercury, a conclusion coming from three investigators. Undoubtedly the fish living where methyl mercury is dumped contain methyl mercury. However, no one has shown this for ocean fish. The method is very difficult, and it is possible that mercury is methylated during the analysis by methyl groups in fish flesh. For example, mercury salts are methylated by acetic acid, which contains a methyl group. If most of the mercury in fish is methyl mercury, then this is a rare example of a non-toxic element being converted to a highly toxic one by natural processes. This type of reaction is anti-evolutionary, and would result in the death of the species.

Therefore, I believe that we can continue to eat fish from everywhere but methyl mercury polluted waters, and that except for this local circumstance, the whole brouhaha is a needless scare. Fish containing much methyl mercury should be taken off the market.

Lead. Children living in tenements may eat lead paint flaking off the old walls, or suck on toys painted with lead (or cadmium) pigment. The result can be mental retardation. In addition to this lead, children playing on streets where there is heavy traffic may absorb considerable lead from leaded gasoline exhausts at ground level. Some studies are beginning to show that such children may be less than average mentally. The Food and Drug Administration has recently limited lead in paint to 600 ppm, which will help successive generations, but not the present.

Other Metals. There are thousands of examples of toxicity from exposure of workers to metals and metal dusts, both overt and innate. Laws prevent airborne contaminants inside plants above certain specified levels, but they are not always obeyed. An example of unknown metallic exposures is the high rates of cancer of the respiratory tract in mechanics and repairmen, foremen, metal and other craftsmen, engine and construction workers, carpenters, public utility workers, painters and plasterers, metal operatives, drivers and deliverymen, manufacturing operatives, and other operatives, but not in professional men, technical men, office workers, miners, guards, managers, clerical workers, salesmen, clerks, laborers, construction workers and outdoor workers. Why? Cancer of the stomach is high in some of these same groups. Why? The only metals that we know cause cancer in workers are nickel, one form of chromium and selenium.

Selenium poisoning occurs in cattle eating grasses and grains grown in areas with lots of natural selenium, and signs of it occur in local people. Marco Polo was the first to describe the disease; hoofs go bad and drop off.

Arsenic poisoning occurred in hundreds of people in Taiwan drinking water from deep wells; we found 71 ppm arsenic in the water. In many of them cancers of the skin developed. The arsenic was natural.

In fact, one can go through the Periodic Table and discover examples of poisoning by each of the elements or nearly all. Too much is too much, as it is with everything but air.

Lithium poisoning can occur when people take too much lithium for manic states, and used to occur when lithium salts were given as substitutes for sodium chloride in low salt diets. Lithium replaces sodium in tissues, and vice versa.

Beryllium disease is found in workers exposed to beryllium dusts, and occasionally in people living in towns where there is much industrial beryllium in the air. They don't get over it. Beryllium, as I have said, is a bad actor. There are about 850 cases recognized, and probably many not recognized.

In certain diseases, too much or too little sodium, potassium, magnesium or calcium in the blood can cause severe symptoms. These changes are secondary to the disease.

Scandium and titanium poisoning are unknown. Vanadium poisoning is confined to workers cleaning petroleum storage tanks, and is due to the direct irritation of the dusts in the lungs. Vanadium workers who absorb a lot get green tongues and little more. A green tongue is an alarming sign, as anyone who has drunk American synthetic crème de menthe can testify when he looks in the mirror the morning after. But knowledge of cause is the cure for anxiety, both in the cases of vanadium and mint.

Chromate workers get cancer of the lungs. Manganese miners get both a form of severe pneumonia and a disease like Parkinsonism, or paralysis agitans, a common disease in old men.

Cobalt added to beer to preserve its foam has caused heart disease in heavy beer drinkers; other, perhaps dietary, factors also are at work. Nickel dusts induce cancers of the nose, sinuses and lungs in exposed workers, and irritation of the skin in sensitive people.

Poisoning by pure zinc is unknown. No examples of

poisoning by gallium or germanium in man are known, nor by aluminum. Selenium can poison industrial workers, who also get a garlic breath. Silicon can collect in the lungs of those who breathe it, causing fibrosis. Babies have been killed by nurses combining boric acid solutions with intravenous fluids. Too much bromide can cause coma.

No cases of poisoning by rubidium, strontium, yttrium, zirconium, niobium, molybdenum, ruthenium, rhodium, or palladium are known. Cadmium poisoning has been discussed. Silver from the frequent use of silver compounds such as Argyrol in the nose or eyes, deposits in the skin of the face, giving a grey color. Silver in the body is very insoluble stuff.

Antimony from enameled ware has entered food and people have been poisoned by the food. Tellurium gives a strong garlic odor to the breath, and in higher doses, suppression of sweating, skin lesions and digestive disturbances in workers.

Tin has a low order of toxicity, fortunately for people who eat canned food, although tin fumes can cause lung disease of a mild nature. People can become sensitive to iodine.

Barium dusts can collect in the lungs of exposed workers, but the disease is mild. No human diseases from lanthanum, hafnium, tantalum, tungsten, rhenium, osmium, iridium, or gold are known. Platinosis has been described from exposure to platinum salts.

Mercury has been discussed. Thallium is very toxic, and has caused serious and fatal poisoning in human beings eating its compounds used as rat-killers, and chronic poisoning from its use as a depilatory. At least 31 Mexicans in California in 1933 died from eating flour from thallium-treated barley intended to kill ground squirrels (which were harboring plague-ridden fleas).

Lead poisoning is not uncommon in workers exposed to paint, fumes and dust. Bismuth poisoning has resulted from its medicinal use.

In local areas, excessive fluoride in drinking water has caused bony changes, but the margin of safety is about 20 to one or more, at the 1 ppm advocated for control of tooth decay, at which level and higher, up to 5 or 10 ppm, no toxicity has appeared.

So curious have physicians always been that many metals have been used to treat various diseases. The list is long: aluminum for silicosis, antimony to induce vomiting (tartar emetic), to treat liver flukes and blood-borne parasites, arsenic for syphilis and other similar infections such as yaws, bismuth for syphilis and receding gums, boric acid as a mild disinfectant, cerium for vomiting of pregnancy, copper for anemia, lithium for mania, gout and arthritis, magnesium as milk of, mercury for syphilis and as a diuretic, palladium for obesity, tin for skin infections, typhoid, osteomyelitis, anthrax and acne, titanium for skin disorders, vanadium for syphilis, tuberculosis, anemia, neurasthenia, and rheumatism (Schroeder's Law of Therapy is that the more diseases a drug claims to cure, the less effective it is), zinc for wound healing (it is effective), and so on.

In general, the essential trace metals are not toxic, and the abnormal ones on the right hand side of the Periodic Table are apt to be toxic.

When a metal or its salts are changed to a synthetic, unnatural form, it can become exceedingly toxic. When an element is combined with hydrogen it becomes very toxic; arsenic to arsine gas, antimony to stibine gas, phosphorus to phosphine gas. When a metal is combined with carbon monoxide to make a metal carbonyl, it can become toxic: nickel or iron carbonyl. For some reason, cobalt carbonyl is not very toxic. When a metal is combined with one to four methyl (CH_3) or ethyl (C_2H_5) groups, it always becomes highly toxic: tetra-ethyl lead, tetra-ethyl tin, methyl or dimethyl mercury, dimethyl cadmium etc. These chemical changes are man-made, like most toxic things.

Thus we see that almost every element in the Periodic Table can be toxic in large enough amounts. Is this a cause for fear?

My teen-aged granddaughter said what many of the young say today, "We are going to kill every living thing on this planet from pollution, and it won't be long." I calmed her fears with knowledge and reason and logic and common sense—I hope, for she was a victim of pollution-paranoia.

There is a simple test one can give oneself which is diagnostic of pollution paranoia. It is based on the free association of one key word with the first one which comes to mind.

Key Word	Normal Answer	Pollution-paranoia Answer
Atomic	power, energy	bomb
Hydrogen	gas	bomb
Strontium	calcium	ninety
Radiation	energy	cancer
Pollution	sewage, smoke	death
Mercury	temperature	poison
Fish	food	mercury
Food	eat	additives
Food Additives	preservation	poisons
Air	fresh	pollution
Water	drink	pollution
Industry	profits	pollution
Automobiles	travel	pollution
Arsenic	old lace	poison
Health	good or bad	health foods
Detergent	clean clothes	eutrophication

There is no National Catastrophe in the offing. There are local areas where things are serious, as there have been and will be. There is less air and water pollution per capita now than a hundred years ago, but there are more people. Our cities have been cleaning their air since 1955, without fanfare—that is, most of them. The laggards must come into line. Our waters are reasonably clean—that is, most of them. The problem is known and things are being done. Our food is too pure; we must make it less so. You can't buy a can of lead paint in Australia without a government license, and the air over Melbourne is clear. London fogs have virtually vanished.

The intelligent human being is good at meeting problems when he knows what they are. But he is liable to go to extremes until brought back into balance. Today we need reason more than anything else, in order to avoid such destructive actions as the ban on swordfish because of supposed methyl mercury content, or the buyers' strike on phosphates (which are good except when put into lakes and rivers in large amounts) or the ban on NTA from hurried, abnormal experiments . . . . To achieve a balanced, reasonable viewpoint, we need knowledge, and most of all, we need to do our homework. Otherwise, we may arrive at the gloomy fizz-out exemplified by the following poem:

POLLUTION

"Do nothing in excess," the wise Greeks said,
　"Collectively or individually;"
In every case, "the piper must be paid."
　—But man ignores his ancient history.
Each culture has within itself the seeds
　Of self-destruction, ruin, and decline.
When heedlessly, to satisfy its needs,
　It flouts the pleas of Nature and of Time.
And so it's come to pass upon this earth:
　Mankind has wrought excess of poverty,
Noise, poisons, hatreds, crimes, sex, Human Birth;
　Excess of everything but Charity.
Thus planets through their own pollution die,
　And float as littered coffins in the sky.

Harry de Metröpolis
(20th Century)

ABOUT THE AUTHOR

HENRY A. SCHROEDER, M.D. is one of this country's most distinguished scientists. His first major contribution was the development of a low-sodium diet, so widely used today for heart conditions. He investigated the causes of hypertension (high blood pressure) from 1937 to 1956, and developed the first effective drug treatment, proving that many severe cases were reversible. As a result, mortality from this cause has been quartered in the past 25 years.

After making experiments on humans for trace elements analyses in the Orient and Middle East, Dr. Schroeder established the metal-free, environmentally controlled Trace Element Laboratory in 1960 on a remote hill in Vermont. In the same year he was vice-chairman of the first World Health Organization symposium held in Prague, Czechoslovakia. His George Brown Memorial Lecture to the American Heart Association in 1966 brought worldwide headlines for its new approaches to the causes of heart disease.

One of three developers of the regulation aviator's anti-blackout suit, Schroeder established standards for aircraft seats and harness to prevent injury from crashes. He also advised on construction of the world's largest human centrifuge, and set standards for the astronauts' space suits. Long before Sputnik he was a consultant on problems to be encountered in space flight.

Dr. Schroeder took his degrees at Yale and Columbia, and holds an honorary M.A. from Dartmouth. He is Professor of Physiology Emeritus at the Dartmouth Medical School, and Director of Research at Brattleboro Memorial Hospital. Born in Short Hills, New Jersey, he has lived in New York City, St. Louis, and Washington, D.C. He has written seven books and hundreds of scientific papers and abstracts.

Index